THE CHEMISTRY
AND BIOCHEMISTRY
OF N-SUBSTITUTED
PORPHYRINS

THE CHEMISTRY AND BIOCHEMISTRY OF N-SUBSTITUTED PORPHYRINS

David K. Lavallee
Department of Chemistry
Hunter College of the
City University of New York

VCH

David K. Lavallee
Department of Chemistry
Hunter College of the
City University of New York
New York, New York 10021

Library of Congress Cataloging-in-Publication Data

Lavallee, David K., 1945-
  The chemistry and biochemistry of N-substituted
porphyrins.

  Bibliography: p.
  Includes indexes.
  1. Porphyrin and porphyrin compounds — Synthesis.
2. Porphyrin and porphyrin compounds — Metabolism.
I. Title.
QP671.P6L38   1987   547.8'69   87-23206
ISBN 0-89573-147-9

Printed in the United States of America.

ISBN 0-89573-147-9 VCH Publishers
ISBN 3-527-266-933 VCH Verlagsgesellschaft

Distributed in North America by:

VCH Publishers, Inc.
220 East 23rd Street, Suite 909
New York, New York 10010-4606

Distributed Worldwide by:

VCH Verlagsgesellschaft mbH
P.O. Box 1260/1280
D-6940 Weinheim
Federal Republic of Germany

ACKNOWLEDGEMENTS *CHEM*

I am very grateful for the editorial assistance and useful suggestions of my in-house editor, Dr. Carolyn Lavallee.

The completion of this project was greatly aided by my opportunity to spend a sabbatical leave as a Fulbright Research Scholar in the laboratory of Dr. Daniel Mansuy. Dr. Mansuy and Dr. Jean-Paul Battioni were gracious hosts and discussions with them were very helpful in developing the chapters on cytochrome P-450 and hydrazines.

I am indebted to several researchers who generously provided me with extensive collections of reprints and preprints: Alan Balch, Henri Callot, Francesco De Matteis, Anthony Jackson, Daniel Mansuy and Paul Ortiz de Montellano. I am grateful to Jean-Paul Battioni, Henri Callot, and James Collman for providing illustrations for the book which had not been published.

Many illustrations have been taken from the literature with the permission of the authors and the following journals: *Arch. Biochem. Biophys., Biochem., Biochem. J., Biochem. Biophys. Res. Comm., Biochem. Pharmacol., Biochim. Biophys. Acta, Inorg. Chem., J. Amer. Chem. Soc., J. Biol. Chem., J. Inorg. Biochem., Molec. Cell. Biol., Molec. Pharmacol.* and *Pure and Appl. Chem.* The source of each of these illustrations is noted within the text.

I gratefully acknowledge the initial support of my research in this field by the Petroleum Research Fund, administered by the American Chemical Society and the continued generous funding by the National Cancer Institute of the National Institutes of Health (grant CA 25427).

CONTENTS

6. Biochemistry of N-Substituted Porphyrins I: Ferrochelatase Inhibition

7. Biochemistry of N-Substituted Porphyrins II: Formation by Cytochrome P-450

8. Biochemistry of N-Substituted Porphyrins III: Formation by Reactions of Hydrazines with Heme Proteins and Migration Reactions of Model Compounds

CHAPTER 1

INTRODUCTION

Porphyrins substituted at a pyrroleninic nitrogen atom (Figure 1.1) are powerful inhibitors of the enzyme ferro-chelatase, which is essential for the formation of heme for hemoglobin, myoglobin and cytochromes. N-substituted porphyrins are produced from the reactions of a number of xenobiotics with the detoxifying cytochrome P-450 enzymes found in the liver: compounds which produce N-substituted porphyrins include prescription drugs, anaesthetics and even such simple compounds as ethylene and acetylene. N-substituted porphyrins also result from the reaction of hydrazines with hemoglobin and myoglobin. The reaction of phenylhydrazine with hemoglobin *in vivo* results in the formation of erythrocyte aggregates called "Heinz bodies" that mimic the capillary-blocking aggregates of malarial patients, which have been known for over a century.

In some cases, the mechanisms of biological reactions of N-substituted porphyrins have been deduced from experiments performed under conditions that closely mimic *in vivo* conditions (using hepatic microsomal preparations). Structural aspects of some naturally-derived N-substituted porphyrins have been deduced directly using NMR spectroscopy. In other cases, model systems have proven essential. For example, all the N-substituted porphyrins whose strucutures have been determined by x-ray crystallography are of synthetic porphyrins. Correlations of spectroscopic properties with structure, used to deduce features of the biological compounds, as well as most mechanisms proposed for biological reactions involving N-substituted porphyrins have been based on studies of synthetic compounds (Figure 1.2). In addition to the important role they have played in understanding biological processes, synthetic N-substituted porphyrins and their complexes are intermediates for rapid synthesis of

radiolabelled metalloporphyrins for use in medical diagnos-
tic imaging[1,2] and are intrinsically interesting because of
their structural novelty.

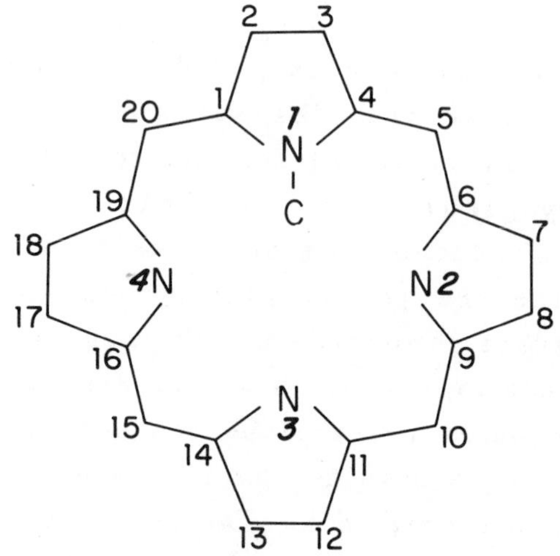

Figure 1.1. The ring structure of an N-substituted porphyrin with the numbering system used throughout this book.
The IUPAC recommended numbering system uses 21, 22, 23 and
24 for the nitrogen atoms. It is gaining adherents,   but
the bulk of current literature uses the system adopted here.

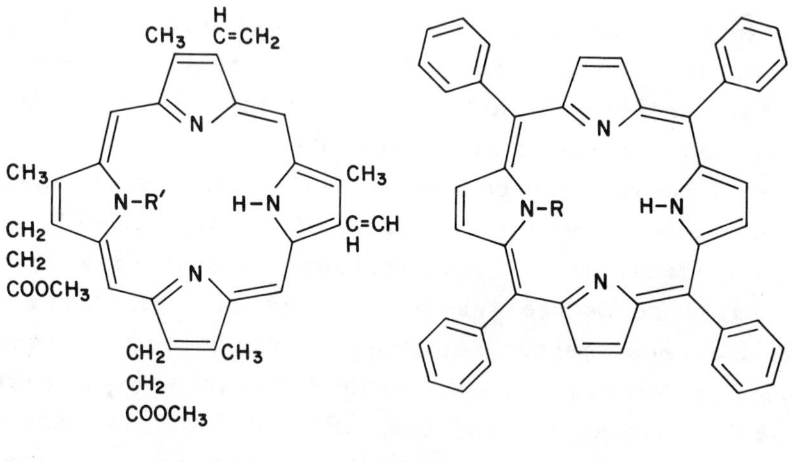

Figure 1.2. The structures of an N-substituted protoporphyrin IX (left) and a *meso*-substituted synthetic porphyrin .

Historical Background. The first report of an N-substituted porphyrin (N-methyletioporphyrin I) was a brief mention in a paper by McEwen in 1936.[3] In 1946, he described the synthesis and characterization of the compound and of its zinc(II) complex.[4] The point of these early studies was to test the effect of distortion on basicity and to see if the N-methylporphyrin could be used as a better endpoint indicator for acid-base titrations than the parent compound, etioporphyrin I. The acid-base properties of N-methylcoproporphyrin were reported by Neuberger and Scott in 1954.[5] In the 1960's, organic chemists interpreting the NMR chemical shifts of protons in aromatic systems had found the concept of "ring current" to be a useful method for estimating the degree of aromatic character of a ring and, of more direct concern with regard to N-substituted porphyrins, of the position of protons relative to the aromatic plane. Since there were no crystallographic data available at the time, the NMR study of N-methyl- and N-ethyletioporphyrin II by Caughey and Iber was the first direct measure of the nonplanarity of an N-substituted porphyrin.[6] Grigg and Johnson and coworkers then directly alkylated metal complexes of corroles and found that N-ethylcarbonyl-methylene porphyrins could be formed from diazoalkanes.[7-9] In 1970, Jackson reported the synthesis and spectral characterization of multiply N-alkylated porphyrins, demonstrating that such multiple substitution at the porphyrin core is not only possible, but facile.[10]

Interest in the effect of distortion of porphyrin planarity then turned to the reactivity of metal complexes, with the studies of relative rates of complexation of etioporphyrin and N-methyletioporphyrin by Hambright's group.[11] In 1970, I began to study N-methyl derivatives of synthetic tetraarylporphyrins with Gebala. Our approach was to systematically characterize the reactivity of metal complexes.[12] Later, Anderson and I systematically investigated the structures of both free base and metal complexes (Chapter 2). My group then investigated relationships between the spectral characteristics and structures of the metal complexes (Chapter 3), their electrochemistry and reactions involving the N-substituent (Chapter 5). Extensive work on the synthesis and reactions of synthetic N-subsituted porphyrin was underway in Callot's laboratory in Strasbourg (Chapters 4 and 5).

In the late 1970's, a number of groups were studying the effects of drugs on heme biosynthesis and on changes in levels of hepatic cytochrome P-450 enzymes. By comparisons of the visible absorption spectra of the anomalous "green pigments" that appeared in the livers of drug-treated rats with those of synthetic non-*meso*-substituted N-alkylporphyrins that had been reported by Jackson and his group,[13] De Matteis and Ortiz de Montellano and their coworkers concluded that the pigments were N-substituted protoporphyrins originating from the prosthetic group of cytochrome P-450 (Chapters 6 and 7). In 1980, Ortiz de Montellano and Kunze demonstrated the production of N-methylprotoporphyrin IX from the liver of rats treated first with phenolbarbitol (to induce cytochrome P-450 production) and then 4-methyldihydrocollidine (DCC)[14] and in 1981 they established the structure of the naturally produced isomer using NMR spectroscopy.[15]

A great deal of work in the area has transpired since these discoveries. By studies of model systems, the groups of Balch, Collman, Dolphin, Mansuy, and Traylor have made important contributions to our understanding of how certain compounds can act as "suicide substrates" during the oxidative processes catalyzed by cytochrome P-450 by their studies of model systems (Chapter 7) and Callot, in particular, has contributed to synthetic, structural and mechanistic information on related metalloporphyrin systems, greatly expanding the chemistry of N-substituted metalloporphyrins (Chapters 2-5, 7, 8). At the same time, Ortiz de Montellano and his group and those of De Matteis, Tephley, and White and others have characterized the products of the enzymatic reaction, in some cases *in vivo* and in others *in vitro*, to determine the requirements for N-alkylprotoporphyrin IX formation and to use this information to learn about the structure and reactions of the enzymatic system (Chapter 7). In addition, the reaction of hydrazines to produce unusual aggregates of red blood cells (Heinz bodies) was studied in synthetic and natural systems by the groups of Mansuy, Callot and Ortiz de Montellano, among others (Chapter 6).

Active research areas include mechanistic studies of model catalytic systems that produce N-alkyated products, experiments designed to provide a deeper understanding of the biochemical pharmacology of cytochrome P-450 reactions, and medical applications of N-substituted porphyrins.

Organization of the Book. This volume is divided into two sections: the chemistry and the biochemistry of N-substituted porphyrins. The chapters of the first section are arranged according to the nature of the chemistry (structural, spectroscopic, general reactions and synthesis) and discussion within the chapters then organized by the type of N-substituent. The biochemical chapters are divided by the class of biochemical reaction and include a brief section at the end concerning the relevant model systems.

References

1. D.K. Lavallee, A. White, A. Diaz, J.-P. Battioni and D. Mansuy, Tetrahedron Letters, **27** (1986) 3521-3524.

2. J. Mercer-Smith, S. Figard, D.K. Lavallee and Z. Svitra, J. Nucl. Med., **26** (1985) 437 and in press (Antibody Mediated Delivery Systems, Marcel-Dekker, New York, 1987).

3. W.K. McEwen, J. Amer. Chem. Soc., **58** (1936) 1124-1129.

4. W.K. McEwen, J. Amer. Chem. Soc., **68** (1946) 711-713.

5. A. Neuberger and J.J. Scott, Proc. Roy. Soc., **A213** (1952) 307-326.

6. W.S. Caughey and P.K. Iber, J. Org. Chem., **28** (1963) 269-270.

7. D.A. Clarke, R. Grigg and A.W. Johnson, <u>J. Chem. Soc.</u>, (1966) 208–209.

8. D.A. Clarke, D. Dolphin, R. Grigg, A.W. Johnson and H.A. Pinnock, <u>J. Chem. Soc.</u>, (1968) 881–885.

9. R. Grigg, A.W. Johnson and G. Shelton, <u>J. Chem. Soc.</u>, (1971) 2287–2294.

10. G.R. Dearden and A.H. Jackson, <u>J. Chem. Soc., Chem. Commun.</u>, (1970) 205–206.

11. B. Shah, B. Shears and P. Hambright, <u>Inorg. Chem.</u>, **10** (1971) 1828–1830.

12. D.K. Lavallee and A.E. Gebala, <u>Inorg. Chem.</u>, **19** (1974) 2004–2008.

13. A.H. Jackson and G.R. Dearden, <u>Ann. New York Acad. Sci.</u>, **206** (1973) 151–176.

14. P.R. Ortiz de Montellano, K.L. Kunze and B.A. Mico, <u>Molec. Pharmacol.</u>, **18** (1980) 602–605.

15. K.L. Kunze and P.R. Ortiz de Montellano, <u>J. Am. Chem. Soc.</u>, **103** (1981) 4225–4230.

CHAPTER 2

CHEMISTRY OF N-SUBSTITUTED PORPHYRINS I:
STRUCTURES OF N-SUBSTITUTED PORPHYRINS AND COMPLEXES

Introduction

As assumed by early workers in this field, addition of a group to one of the central nitrogen atoms distorts the normally-planar porphyrin ring. The degree of this distortion and the resultant effects on the bonds throughout the porphyrin ring can only be determined by x-ray crystallography. The systematic structural changes brought about by different metal ions is also only accessible through crystallographic studies. In this chapter, the major parameters by which N-substituted porphyrins and their complexes differ structurally from non-N-substituted porphyrins which are emphasized are: the relative orientations of the pyrrolenine rings with respect to the plane of the central nitrogen atoms (a parameter which indicates the degree of "ruffling" of the porphyrin ring), changes in bond lengths in the substituted pyrrolenine ring (indicating how the N-substituent changes the π electron delocalization pattern – also evident in the NMR spectra discussed in Chapter 3), changes in hybridization at the substituted nitrogen atom (especially important for metal complexes in which the degree of interaction of the metal atom with the nitrogen atom can vary considerably), the out-of-plane distance of the metal atom, and the differences in bond angles that correspond to each of these bond length differences.

Although relatively few structures of N-substituted porphyrins have been determined by x-ray crystallographic methods, the trends in structural parameters of these few indicate that the topology of the porphyrin ring and the geometry of the coordination site are highly predictable.

That a consistent pattern has been observed in these structures is important because the only type of ligands for
which there are crystallographic data are synthetic porphyrins (*meso*-tetraphenyl-, octaethyl-, and etioporphyrins)
rather than the naturally derived porphyrins of biochemical
interest.   In addition, spectral and structural parameters
correlate well, indicating that spectral parameters alone
may provide an important means of predicting the structures
of numerous additional N-substituted porphyrins.

Structures of two types of N-substituted porphyrins have
been determined crystallographically: those in which the
substituent is bound to a single nitrogen of the porphyrin
coordination site (herein referred to simply as N-alkyl- or
N-arylporphyrins) and those in which the substituent forms
two bonds to the porphyrin – either to two porphyrin nitrogen atoms or to one nitrogen atom and a metal atom.

Crystal Data. The N-substituted derivatives of tetraphenylporphyrin are relatively easy to crystallize and many
are highly stable. A series of structures of these species
has been reported, including a free base,[1] N-methyl-
5,10,15,20-tetrakis(p-bromophenyl)porphyrin (N-CH$_3$TPPBr$_4$)
and several metal complexes of N-methyl-5,10,15,20-tetra-
phenylporphyrin[2-6] (Mn(N-CH$_3$TPP)Cl, Fe(N-CH$_3$TPP)Cl, Co(N-
CH$_3$TPP)Cl, and Zn(N-CH$_3$TPP)Cl), an N-benzylporphyrin
complex[7] (Zn(N-bzTPP)Cl) and an N-arylporphyrin complex[8]
(Zn(N-phTPP)Cl). Structures of two N-ethylcarbonylmethyl-
2,3,7,8,12,13,17,18-octaethylporphyrins, the monocation[9] (N-
CH$_2$COC$_2$H$_5$H$_2$OEP$^+$ I$^-$) and a cobalt complex[10] (Co(N-
CH$_2$COC$_2$H$_5$OEP)Cl), have also been reported. The crystal data
for all these species are given in Table 2.1. In addition,
one structure of a di-N-alkylated porphyrin, N, N'-dimethyl-
etioporphyrin I monocation (I$_3^-$ salt) is known.[11]   The
structures of the M(N-CH$_3$TPP)Cl complexes are isomorphous,
with packing diagrams similar to that for Fe(N-CH$_3$TPP)Cl
shown in Figure 2.4. Of these complexes, the best agreement
factor was obtained for the Fe(II) complex and its structural parameters will be discussed in detail. For the other

complexes of the series the significance level of the bond lengths is on the order of 0.01A and for the bond angles 1°.

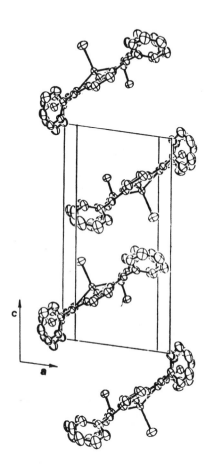

Figure 2.1. The packing diagram for chloro-N-methyl-5,10,15,20-tetraphenylporphinatoiron(II).[3] The nature of the packing in this crystal and those of other N-methyltetraphenylporphyrin complexes indicates that no major structural feature of the molecule is determined by crystal packing forces.

Table 2.1.   Crystal Data for N-Substituted Porphyrins

Free Base N-Alkylporphyrin:
N-methyl-5,10,15,20-tetrakis(p-bromophenyl)porphine[a]

$P2_{1/c}$ (Z=4), CuKα, 3860 I>2.5 (I), R=0.082, R_w=0.098
a=15.440(2), b=16.261(2), c=17.534(2)Å
V=4183.0Å^3, ρ=1.64 g/cm^3

Monoprotonated N-Alkylporphyrin:
21-ethylcarbonylmethyl-2,3,7,8,12,13,17,18-octaethyl-
    porphinium iodide[b]

$P\bar{1}$ (Z=2), MoKα, 2145 I>3 (I), R=0.071
a=14.485(2), b=14.957(2), c=10.360(1)Å
α=92.81(6), β=101.36(6), γ =103.68(5)°
V=2127.3Å^3, ρ=1.26 g/cm^3

Mn(II) Complex of an N-methylporphyrin:
Chloro-N-methyl-5,10,15,20-tetraphenylporphinatomanganese(II)[c]

$P\bar{1}$ (Z=2), MoKα, 3815 I>3 (I), R=0.072, R_w=0.094
a=7.558(6), b=14.993(12), c=14.476(13)Å
α=103.13(2), β=97.09(3), σ =94.00(2)°
V=1903.9Å^3, ρ=1.25 g/cm^3

Fe(II) Complex of an N-methylporphyrin:
Chloro-N-methyl,5,10,15,20-tetraphenylporphinatoiron(II)[d]

$P\bar{1}$ (Z=2), MoKα, 4056 I>3 (I), R=0.043, R_w=0.055
a= 7.5620(6), b=14.961(2), c=17.434(2)Å
α=103.15(1), β=97.124(9), γ =93.84(1)°
V=1896.6Å^3, ρ=1.26 g/cm^3

Table 2.1. Continued

Co(II) Complex of an N-methylporphyrin:
Chloro-N-methyl-5,10,15,20-tetraphenylporphinatocobalt(II)[e]

$P\bar{1}$ (Z=2), MoKα, 3703 I>3 (I), R=0.067, R_w=0.087
a=7.484(2), b=14.972(4), c=17.398(5)Å
α=102.93(1)°, β=97.03(1)°, γ=94.13(1)°
V=1875.7Å^3, ρ=1.28 g/cm^3

Co(II) Complex of an N-alkylporphyrin:
Chloro-N-ethylcarbonylmethyl-2,3-7,8,12,13,17,18-octaethyl-
porphinatocobalt(II)[f]

$P2_{1/n}$ (Z=4), MoKα, 2417 I>3 (I), R=0.041
a=15.214(4), b=17.020(7), c=14.760(2)Å
β=93.59°
V=3814.5Å^3, ρ=1.24 g/cm^3

Zn(II) Complex of an N-methylporphyrin:
Chloro-N-methyl-5,10,15,20-tetraphenylporphinatozinc(II)[g]

$P\bar{1}$ (Z=2), MoKα, 3302 I>3 (I), R=0.069, R_w=0.076
a=11.970(6), b=13.468(8), c=14.998(7)Å
α=101.73(1)°, β=107.00(2)°, γ=115.88(2)°
V=1920.4Å^3, ρ=1.36 g/cm^3

Zn(II) Complex of an N-arylporphyrin:
Chloro-N-phenyl-5,10,15,20-tetraphenylporphinatozinc(II)[h]

$P2_12_12_1$ (Z=4), MoKα, 2928 I>2 (I), R=0.055, R_w=0.057
a=15.078(5), b=15.272(5), c=17.268(5)Å
V=3976.3Å^3

---

a Ref. 1, b Ref. 9, c Ref. 2, d Ref. 3, e Ref. 4,
f Ref. 10, g Ref. 6, h Ref. 8.

Structures of Metal-Free N-Substituted Porphyrins

N-alkylation produces greater structural changes within a pyrrolenine ring of the porphyrin than does protonation. These changes would be consistent with a greater degree of double bond character in the $C_a-C_b-C_b-C_a$ portion of the ring and greater π delocalization toward the ring periphery rather than toward the nitrogen atoms. Bond lengths for the porphyrin ring systems of N-CH_3TPPBr$_4$ and of N-$CH_2COC_2H_5H_2OEP^+$ I^- are shown in Figure 2.2. The data in parentheses are for corresponding structures without N-substituents, H_2TPP and H_3OEP^+, respectively. In both cases, the bond lengths in the alkylated pyrrolenine ring are different from those in the other three rings, but merely at the significance level. The pattern is: longer N-C_a bonds (by about 0.05A), shorter C_a-C_b bonds (by about 0.05A) and a longer C_b-C_b bond for the alkylated ring. In the non-N-alkylated structures, the C_b-C_b bond lengths are intermediate between those of the N-alkylated and non-N-alkylated rings of the N-alkylporphyrins, as might be expected. The bond lengths of the protonated pyrrole rings of the N-alkylporphyrins match the protonated rings of the non-N-alkylated porphyrins.

The bond angles for N-CH_3TPPBr$_4$ and N-$CH_2COC_2H_5H_2OEP^+$ (with data for H_2TPP and H_2OEP shown in parentheses, Figure 2.3) provide only tentative indications of trends within the pyrrole rings. The angular parameter of greatest interest is the angle between the nitrogen-substituent carbon bond and the plane of the alkylated pyrrole ring. For N-CH_3TPPBr$_4$, this angle is $120°$ and for N-$CH_2COC_2H_5H_2OEP^+$ it is $115°$. For a protonated pyrrole (sp^3) the corresponding angle is $180°$ while for sp^3 hybridized nitrogen it would, of course, be $109°$. As noted earlier, the N-C_a bond length changes parallel a decrease in bond order that accompanies a change from sp^3 toward sp^3 hybridization.

Another aspect of N-alkylporphyrin structures is the relative rotation of the pyrrole rings about the C_a-C_m bonds. Although some of the non-N-alkylated porphyrins exhibit distortion of this kind in the solid state (often referred to as "ruffling"), the deformations observed for the N-alkylpor-phyrins are more pronounced and, in contrast to the non-N-alkylated free base porphyrins, cannot be attributed to lattice packing energetics. As Table 2.2 shows, the N-alkylated ring is rotated to a great extent. In the case of the free-base N-methyl porphyrin, the protonated ring, located opposite the alkylated ring, is rotated to a lesser extent but in the same direction. In the case of the mono-cation of the N-substituted octaethylporphyrin, the two pro-tonated rings are located adjacent to the alkylated ring and are rotated in the opposite sense. For all of the metal complexes the three nonalkylated rings are bound to the metal atom and, as a result, are rotated so that the lone pair of the nitrogen atom is oriented toward the metal atom and in the opposite sense to the alkylated ring. The alkylated pyrrole ring must be rotated in this manner to prevent contact of the alkyl group and the metal atom.

The structure of N,N'-dimethyletioporphyrin I monocation I_3^- salt is unique - the sole example of a multiply alkylated porphyrin. Since etioporphyrin I is symmetric, with alternating methyl and ethyl groups on the β-carbon atoms of the pyrrolenine rings, there is only one geometric isomer for the *cis*-methyl derivative. In the structure reported by Grigg and coworkers,[11] the two methyl groups are *anti* with respect to the porphyrin ring. This configuration is also called *trans*, which can lead to some confusion with the designation of *trans* sometimes used to denote which nitrogen atoms are alkylated. The neighboring unsubstituted pyrrolenine rings which bear the methyl groups are canted at angles of 37° and -32° with respect to the plane of the four nitrogen atoms while the non-N-substituted rings are canted at angles of only 4° and 1°, much like the angles found for mono-N-substituted porphyrins. It is interesting that in this structure the C_b-C_b distances of the alkylated rings are not much longer than the corresponding bond in the non-substituted rings (the values in Å are 1.374 and 1.329 for the N-methylated rings and 1.364 and 1.297 for the non-N-methylated rings). The authors proposed a delocalization pattern which isolates the C_b–C_b bonds of the two opposite rings (one N-methylated, at 1.329 Å, and one ring that has neither an N-methyl group nor N-H, at 1.297 Å) to accommodate the bond length change.[11] The N-C distances (1.496(13) Å and 1.481(12) Å) are typical of those for other N-alkylporphyrins. The other bond distances and angles are also quite consistent with those observed for the mono-N-methyl porphyrin and the N-ethylcarbonylmethyl porphyrin monocation structures.

Table 2.2 Deviations From the Least-Squares Plane of the Unsubstituted Nitrogen Atoms[a]

	N-substituted Ring	Opposed Ring	Adjacent Rings	Out-of-Plane Displacement
Free Base, N-CH$_3$TPPBr$_4$	27.7°	8.1°	−11.9°, −10.2°	N.A.
Monoprotonated, N-CH$_2$COC$_2$H$_5$HOEP$^+$	19.1°	−11.7°	−4.8°, −2.2°	N.A.
Mn(II) Complex, Mn(N-CH$_3$TPP)Cl	29.2°	−4.2°	−12.0°, −9.8°	0.69Å
Fe(II) Complex, Fe(N-CH$_3$TPP)Cl	36.6°	−6.4°	−11.3°, −9.8°	0.62Å
Co(II) Complex, Co(N-CH$_3$TPP)Cl	31.6°	−5.9°	−10.4°, −8.3°	0.56Å
Co(II) Complex, Co(N-CH$_2$COC$_2$H$_5$OEP)Cl	44.1°	−10.8°	−12.2°, −8.3°	--
Zn(II) Complex, Zn(N-CH$_3$TPP)Cl	38.5°	−5.3°	−12.3°, −11.6°	0.65Å
Zn(II) Complex, Zn(N-PhTPP)Cl	42.0°			0.67Å

<u>a</u> For references, see Table 2.1.

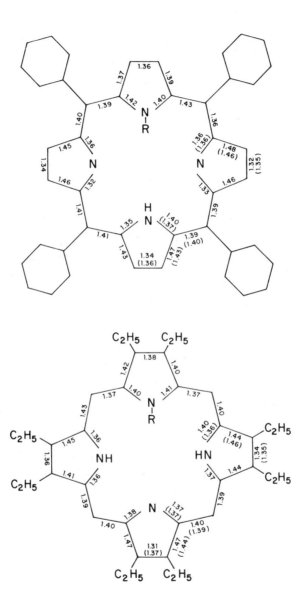

Figure 2.2. Bond lengths (Å) for the structures of N-methyl-5,10,15-20-tetrakis(p-bromophenyl)tetraphenylporphine[1] and N-ethylcarbonylmethyl-2,3,7,8,12,13,17,18-octaethylporphyrin monocation.[12]

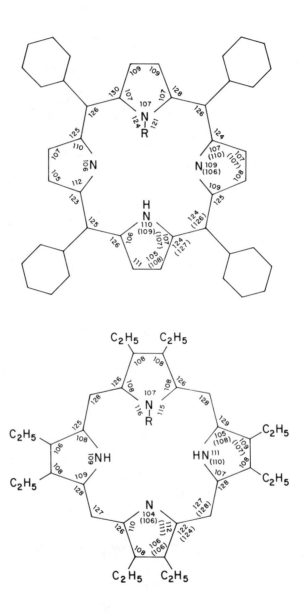

Figure 2.3. Bond angles (degrees) for N-methyl-5,10-15,20-tetrakis(p-bromo)phenylporphine[1] and N-ethylcarbonylmethyl-2,3,7,8,12,13,17,18-octaethylporphyrin monocation.[12]

Structures of N-Substituted Metalloporphyrins

Among the major differences in the structures of metal
complexes of N-substituted and non-N-substituted porphyrins
are the unique bond to the substituted nitrogen atom and the
out-of-plane orientation of the metal atom. Many structures
of non-N-substituted metalloporphyrins which are five coordinate also have the metal atom out-of-plane, but in those
cases, the four metal-nitrogen bond lengths are either the
same or very similar.[12] In discussing the nature of the
out-of-plane metal atom, therefore, the reference plane is
generally taken to be the four pyrrolenine nitrogen atoms
for non-N-substituted complexes but for N-substituted complexes it is typically taken to be the plane formed by the
three unsubstituted nitrogen atoms. The structure of
chloro-N-methyl-5,10,15,20-tetraphenylporphinatoiron(II),
(Fe(N-CH$_3$TPP)Cl), is typical of those for N-substituted
metalloporphyrins.

The bond lengths and angles of Fe(N-CH$_3$TPP)Cl are shown
in Figure 2.4 with values for a corresponding non-N-alkylated high spin iron(II) porphyrin,[13] Fe(TPP)(2-MeIm), shown
in parentheses. The coordination geometry about the metal
atom is best described as a distorted square-based pyramid,
with the four nitrogen atoms of the N-methylporphyrin ligand
occupying the basal sites and the chloro ligand in the
apical position making a strong bond to the iron(II) atom
(Fe-Cl = 2.244(1) Å). The rehybridization of the alkylated
porphyrin nitrogen atom and the bulk of the N-methyl group
lead to a large displacement of the metal ion from the
reference plane of the three nonalkylated nitrogen atoms and
toward the apical chloro ligand. The 0.62 Å apical
displacement of the iron(II) atom is comparable to the those
of other M(N-CH$_3$TPP)Cl complexes (Table 2.2). This apical
displacement is considerably larger than the distances of
0.42 Å and 0.40 Å seen in the nonalkylated five-coordinate
high-spin Fe(II) complexes Fe(TPP)(2-MeIm) and Fe(TPivPP)(2-
MeIm) (where 2-MeIm is 2-methylimidazole and TPivPP is
5,10,15,20-tetrakis(o-pivalamidophenylporphyrin).[14,15]

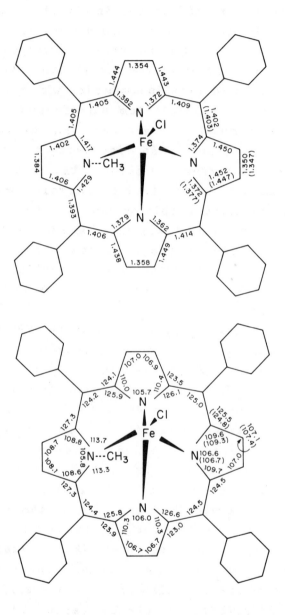

Figure 2.4. Bond lengths and angles of chloro-N-methyl-5,10,15,20-tetraphenylporphinatoiron(II)[3] with values for 2-methylimidazoyl-5,10,15,20-tetraphenylporphinatoiron(II)[13] in parenthesis.

As shown in Table 2.3, the metal–nitrogen bonds for the four M(N–CH$_3$TPP)Cl complexes follow a very similar pattern. In each case, only one metal–nitrogen bond is as strong and short as in comparable planar porphyrin complexes. This is the bond to the nitrogen atom N3, *trans* to the alkylated nitrogen atom (N1). For example, the Fe–N3 distance of 2.082 Å is very similar to the Fe–N distances of 2.068–2.092 Å reported for the Fe(TPP)(2–MeIm) and Fe(TPivPP)(2–MeIm).[13,15] The other two nonalkylated nitrogen atoms form slightly longer bonds to the iron atom (Fe–N4 = 2.118 Å, Fe–N2 = 2.116 Å), just as found for the corresponding Mn(II), Co(II) and Zn(II) complexes. All these are longer than the Fe(II)–N distance of 1.972 Å in four-coordinate Fe(TPP).[16]

The interaction between the iron(II) atom and the methylated nitrogen atom (N1) is much weaker than the other Fe–N bonds, as shown by the Fe–N1 distance of 2.329 Å in Fe(N–CH$_3$TPP)Cl. The shorter Fe–N distances relative to Mn–N distances are due to the smaller radius of the Fe(II) atom. This trend is found for the strong metal–nitrogen bonds (M–N4, M–N2, and M–N3) of the Co(N–CH$_3$TPP)Cl complex, although the tendency of the cobalt(II) ion for four coordination leads to a lengthening of the weaker Co–N1 bond relative to the Fe(II) and Mn(II) cases. The tendency for four coordination is maximum in the Zn(II) complex, where the Zn–N1 distance is roughly 0.2 Å longer than the Fe–N1 distance. Another way of viewing the trends in metal–nitrogen bond distances of the M(N–CH$_3$TPP)Cl complexes is that the strong bonds determine the structure so that a smaller metal atom is drawn closer to N3. The M–N1 bond is then lengthened by default. The weakening of the M–N1 bond is reflected in the N 1s energies determined by x-ray photoelectron experiments discussed later. Weakening of the M–N1 bond is also reflected in the N1–M–Cl angle, which is about 104° in the two cases where the M–N1 interaction is strongest [Mn(II), Fe(II)] and drops to approximately 95° for the metal atoms for which the interaction is less important [Co(II), Zn(II)]. Further similarities in the coordination of the four M(N–CH$_3$TPP)Cl complexes are evident in Table 2.3.

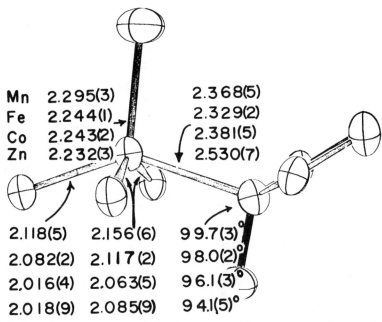

Mn	2.295(3)	2.368(5)
Fe	2.244(1)	2.329(2)
Co	2.243(2)	2.381(5)
Zn	2.232(3)	2.530(7)

2.118(5)	2.156(6)	99.7(3)°
2.082(2)	2.117(2)	98.0(2)°
2.016(4)	2.063(5)	96.1(3)°
2.018(9)	2.085(9)	94.1(5)°

Figure 2.5. Parameters of the coordination sphere for a series of M(N–CH$_3$TPP) complexes.[3]

Table 2.3. Metal–Ligand Distances (Å) and Angles (Deg) in M(N–CH$_3$TPP)Cl Complexes[a]

	Mn(II)	Fe(II)	Co(II)	Zn(II)
			M(II)	
M–Cl	2.295(3)	2.244(1)	2.243(2)	2.232(3)
M–N1	2.368(5)	2.329(2)	2.381(5)	2.530(7)
M–N2	2.156(5)	2.116(2)	2.063(5)	2.081(9)
M–N3	2.118(5)	2.082(2)	2.016(2)	2.018(9)
M–N4	2.155(6)	2.118(2)	2.063(5)	2.089(6)
N1–M–Cl	104.2(1)	103.48(6)	96.7(1)	94.5(2)
N2–M–Cl	109.6(2)	108.03(7)	105.9(2)	105.6(2)
N3–M–Cl	118.6(2)	114.38(7)	116.6(2)	120.8(2)
N4–M–Cl	106.4(1)	105.36(7)	103.3(1)	107.4(2)
M–N1–C1	110.9(4)	112.7(2)	112.7(4)	107.9(4)
M–N1–C4	111.4(4)	113.5(2)	113.6(4)	108.8(5)
M–N1–C21	99.7(3)	98.0(2)	96.1(3)	94.1(5)
C1–N1–C4	105.8(5)	105.8(2)	106.3(5)	108.2(8)
C1–N1–C21	114.3(5)	113.3(2)	114.1(5)	118.5(8)
C4–N1–C21	114.9(5)	113.7(2)	114.1(5)	117.7(6)

a Ref. 3, estimated standard deviations in parentheses

Table 2.4. Delocalization Patterns in Metalloporphyrins $(\overset{\circ}{A})^{\underline{a}}$

complex	$N–C_a$	$C_a–C_b$	$C_b–C_b$
	nonmethylated pyrrole rings (methylated pyrrole rings)		
non–N–substituted[b]	1.379(6)	1.443(5)	1.354(10)
Mn(N–CH₃TPP)Cl	1.373(7)	1.448(9)	1.356(9)
	(1.424(8))	(1.406(8))	(1.398(10))
Fe(N–CH₃TPP)Cl	1.374(3)	1.446(4)	1.354(4)
	(1.423(3))	(1.404(4))	(1.384(4))
Co(N–CH₃TPP)Cl	1.381(8)	1.448(9)	1.356(9)
	(1.416(8))	(1.410(9))	(1.382(9))
Zn(N–CH₃TPP)Cl	1.38(1)	1.44(1)	1.34(2)
	(1.40(1))	(1.42(2))	(1.35(1))

<u>a</u> Ref. 3.
<u>b</u> Average of data for 16 complexes (from ref. 13).

The entries in Table 2.4 are designed to compare the bond lengths, and thus the effective delocalization patterns, within the pyrrole rings of the N-methylporphyrin ligands. In general, the bond lengths within the nonmethylated pyrrolenine rings similar to those of non–N–alkylated porphyrin complexes and to those of $HN–CH_3TPPBr_4$ discussed previously. As we saw for the free base N-methylporphyrin, the N–C and C–C distances of the methylated pyrrole ring are quite different from those of the corresponding non–N–alkylated species, as a result of the change in geometry about the nitrogen atom upon methylation. In these metal complexes as well as in the free ligand, the shift toward sp^3 character upon methylation results in $N–C_a$ bonds which are distinctly longer than the normal $N–C_a$ bonds of non–N–alkylated porphyrins or of the non–methylated pyrrolenine rings of these complexes. At the same time, the $C_a–C_b$ bonds for the methylated rings are significantly shorter in each case than the corresponding bonds in the nonmethylated pyrrole rings while the C_b-C_b bonds are generally longer in the methylated pyrrole ring. Thus the change in hybridization about N1 and the concomitant large tilt of the methylated pyrrole ring relative to the N2-4 reference plane (Table 2.2, approximately 30°) has a slight but significant change in the delocalization in the methylated pyrrole ring. This extent of this change in delocalization is similar for both the free ligand and metal complexes.

Table. 2.5. Bond Lengths (Å) and Angles (deg) Involving Co(II) In N-Methylcarbonylethyl- and N-Methylporphyrin Complexes

	N-CH$_2$COC$_2$H$_5$OEP Complex[a]	N-CH$_3$TPP Complex[b]		N-CH$_2$COC$_2$H$_5$OEP Complex	N-CH$_3$TPP Complex
		Bond Lengths[c]			
Co(II)-N1	2.455(5)	2.381(5)	Co(II)-N4	2.063(5)	2.063(5)
Co(II)-N2	2.072(5)	2.063(5)	Co(II)-Cl	2.271(2)	2.243(2)
Co(II)-N3	1.992(5)	2.016(4)			
		Bond Angles			
N1-Co-Cl	91.2(1)	96.7(1)	Co-N1-C1	109.5(2)	112.7(4)
N2-Co-Cl	109.1(1)	105.9(2)	Co-N1-C4	109.2(2)	113.6(4)
N3-Co-Cl	119.0(1)	116.6(2)	Co-N1-C21	95.3(2)	96.1(3)
N4-Co-Cl	102.7(1)	103.3(1)	Co-N2-C6	129.4(3)	130.9(3)
N1-Co-N2	81.2(2)	81.5(2)	Co-N2-C9	122.8(2)	123.3(4)
N1-Co-N3	149.7(2)	146.7(2)	Co-N3-C11	126.5(3)	125.8(4)
N1-Co-N4	81.4(2)	81.9(2)	Co-N3-C14	127.1(2)	126.4(4)
N2-Co-N3	89.8(2)	89.6(2)	Co-N4-C16	122.2(2)	123.6(4)
N2-Co-N4	143.9(2)	147.8(2)	Co-N4-C19	128.8(2)	130.0(4)
N3-Co-N4	89.5(2)	89.4(2)			

[a] Ref. 10, [b] Ref. 5, [c] Est. Stan. Dev. in Parentheses.

Table 2.6. Bond Lengths(Å) N-Substituted Porphyrin Ligands in Chlorocobalt(II) Complexes

	N-CH$_2$COC$_2$H$_5$OEP[a]	N-CH$_3$TPP[b]		N-CH$_2$COC$_2$H$_5$OEP	N-CH$_3$TPP
N1-C21	1.456(8)	1.504(8)	N3-C11	1.376(8)	1.373(7)
N1-C1	1.411(8)	1.415(7)	N3-C14	1.375(7)	1.377(9)
N1-C4	1.427(8)	1.416(9)	C10-C11	1.391(9)	1.391(9)
C1-C2	1.409(9)	1.415(9)	C11-C12	1.443(8)	1.459(9)
C2-C3	1.386(8)	1.382(9)	C12-C13	1.363(9)	1.355(10)
C3-C4	1.404(9)	1.405(8)	C13-C14	1.445(8)	1.443(8)
C4-C5	1.383(9)	1.393(8)	C14-C15	1.378(8)	1.392(8)
C1-C20	1.383(9)	1.389(9)	N4-C16	1.375(7)	1.374(8)
N2-C6	1.385(7)	1.387(8)	N4-C19	1.376(7)	1.388(7)
N2-C9	1.365(7)	1.385(7)	C15-C16	1.366(8)	1.397(8)
C5-C6	1.368(9)	1.387(8)	C16-C17	1.469(8)	1.452(9)
C6-C7	1.455(8)	1.454(9)	C17-C18	1.332(8)	1.360(10)
C7-C8	1.330(9)	1.353(9)	C18-C19	1.463(8)	1.444(10)
C8-C9	1.458(9)	1.434(10)	C19-C20	1.379(8)	1.395(9)
C9-C10	1.393(8)	1.413(9)			

Table 2.7. Bond Angles(deg) for N-Substituted Porphyrin Ligands in Chlorocobalt(II) Complexes

	N-CH$_2$COC$_2$H$_5$OEP[a]	N-CH$_3$TPP[b]		N-CH$_2$COC$_2$H$_5$OEP	N-CH$_3$TPP
C1-N1-C21	117.4(4)	114.1(5)	C9-C10-C11	126.6(4)	125.4(5)
C4-N1-C21	117.8(4)	114.1(5)	C10-C11-C12	126.6(4)	125.9(5)
N1-C1-C2	108.8(4)	108.6(6)	C10-C11-N3	123.3(4)	125.1(5)
N1-C1-C20	122.1(4)	124.1(5)	N3-C11-C12	110.5(4)	109.0(6)
C20-C1-C2	128.7(4)	127.2(5)	C11-N3-C14	106.1(4)	106.7(5)
C1-N1-C4	106.3(4)	106.3(5)	C11-C12-C13	106.2(4)	107.4(5)
C1-C2-C3	108.1(4)	107.8(5)	C12-C13-C14	107.4(4)	106.7(6)
C2-C3-C4	108.5(5)	108.7(6)	C13-C14-N3	109.8(4)	110.2(5)
C3-C4-N1	108.3(4)	108.6(5)	N3-C14-C15	122.3(4)	124.3(5)
N1-C4-C5	120.8(4)	123.5(5)	C13-C14-C15	127.8(4)	125.4(6)
C3-C4-C5	130.6(4)	127.9(6)	C14-C15-C16	127.9(4)	125.4(6)
C4-C5-C6	126.8(4)	120.0(5)	C15-C16-C17	124.7(4)	123.1(6)
C5-C6-C7	125.7(4)	123.8(6)	C15-C16-N4	126.2(4)	125.5(6)
C5-C6-N2	125.2(4)	126.1(5)	C16-N4-C19	106.5(4)	105.3(5)
N2-C6-C7	108.8(4)	110.0(5)	C16-C17-N4	109.1(4)	110.3(5)
C6-N2-C9	106.5(4)	105.2(5)	C16-C17-C18	107.9(4)	107.2(7)
C6-C7-C8	107.7(4)	106.7(6)	C17-C18-C19	106.8(5)	106.4(6)
C7-C8-C9	107.2(5)	107.5(5)	N4-C19-C20	126.2(4)	125.8(6)
C8-C9-N2	109.6(4)	110.6(5)	C18-C19-N4	109.6(4)	110.7(5)
C8-C9-C10	123.8(4)	123.0(5)	C18-C19-C20	123.7(4)	123.4(5)
N2-C9-C10	126.6(4)	126.4(6)	C19-C20-C1	126.8(4)	124.6(5)

[a] Ref. 10, [b] Ref. 5, [c] Est. stan. dev. in parentheses.

The other structures of N-substituted complexes that have reported are those of chloro-N-ethylcarbonylmethyl-2,3,7,8,12,13,17,18-octaethylporphinatocobalt(II)[10] (Co(N-CH$_2$COC$_2$H$_5$OEP)Cl, chloro-N-phenyl-5,10,15,20-tetraphenylporphinatozinc(II)[8] (Zn(N-PhTPP)Cl) and chloro-N-benzyl-5,10,15,20-tetraphenylporphinatozinc(II)[7] (Zn(NBzTPP)Cl). The major structural features of the complexes Co(N-CH$_2$COC$_2$H$_5$OEP)Cl and Co(N-CH$_3$TPP)Cl are very similar. The differences in the structural parameters for these two complexes, shown in Tables 2.5, 2.6 and 2.7, are generally due to inherent differences between the N-substituted OEP and TPP ligands (Figure 2.3). There are, however, some significant differences in the cobalt(II) coordination spheres. Both the cobalt-to-substituted nitrogen atom and cobalt-to-chloride bond distances are shorter in the N-CH$_3$TPP complex (by 0.07 and 0.03 Å, respectively) and bond angles those involving the chloro ligand and the cross-ring angles N1-Co-N3 and N2-Co-N4 differ by several degrees. At present there are no other structural data of N-substituted complexes of the same metal but different porphyrins to allow these differences to be confidently attributed to the difference in peripheral substituents. The very similar structures[6-8] of Zn(N-CH$_3$TPP)Cl, Zn(N-PhTPP)Cl, and Zn(NBzTPP)Cl indicate the N-substituent probably has less influence on the structure than the differences between Co(NCH$_3$TPP)Cl and Co(N-CH$_2$COC$_2$H$_5$OEP).

The structures of three complexes which differ only in the N-substituent, Zn(N-phTPP)Cl, Zn(N-CH$_3$TPP)Cl, and Zn(N-bzTPP)Cl, illustrate a relationship between ground state structures and reactivity. The tendency for an N-substituent to be removed by a nucleophile from these complexes is directly related to the ability of the N-substituent to stabilize the incipient carbocation,[17] so that the reactivity order is benzyl, methyl, phenyl. The Zn(II) atom is closer to the plane of the nitrogen atoms and its coordination is less tetrahedral for the more reactive complex (Table 2.8).

Table 2.8. Structural Parameters of Chlorozinc(II) N-benzyl, N-methyl- and N-phenyl- Tetraphenylporphyrins.[a]

	Zn(N-PhTPP)Cl	Zn(N-CH$_3$TPP)Cl	Zn(N-BzTPP)Cl
Zn–Cl	2.224(2)	2.232(3)	2.265(1)
Zn–N1	2.490(5)	2.530(7)	2.477(3)
Zn–N2	2.093(5)	2.089(6)	2.080(3)
Zn–N3	2.016(5)	2.018(9)	2.023(3)
Zn–N4	2.107(5)	2.081(9)	2.092(3)
Metal above plane[b]	0.67	0.65	0.59
N1–Zn–Cl	97.0(5)°	94.5(2)°	93.2(1)°
Cant of pyrrole[b]	42°	39°	32°

[a] Ref. 7, distances in Ångstroms.
[b] Relative to the N2, N3, N4 reference plane.

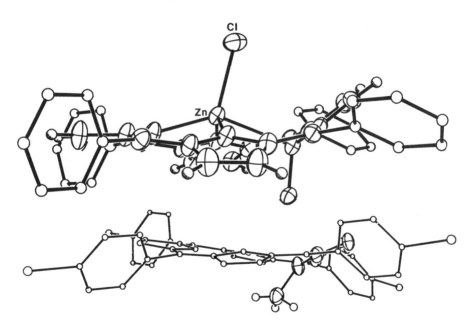

Figure 2.6. Side-views of N-methyl-5,10-15,20-tetrakis(p--bromophenyl)porphine[1] and chloro-N-methyltetraphenyl-porphinatoiron(II),[3] showing the similarity of HN–CH$_3$TPP and its metal complexes.

In addition to the three structures of zinc complexes
which only differ by the N-substituent, there is now a
second structure of an iron complex which can be compared
with that of Fe(N-CH$_3$TPP)Cl. In order to settle a contro-
versy concerning the nature of intermediates (whether carbene or N-vinylidene complexes) formed during reactions of
DDT with iron porphyrins,[18] Balch and coworkers determined
the structure of chloro-2,2-bis(p-chlorophenyl)-N-vinyl-
tetraphenylporphinatoiron(II).[20] Table 2.9 provides a
comparison of the bond lengths and angles of the chloro-
iron(II) complexes of these N-methyl- and N-vinylporphyrins
and, as well, parameters for a carbene complex derived from
a close analog of DDT (only lacking the p-chloro substituents on the two phenyl rings, the structure of this
carbene complex is such that the carbon atom is inserted
between the iron atom and a pyroleninic nitrogen atom).
Certainly, these data clearly demonstrate the closer
similarity of the N-vinylidene complex with the N-methyl
complex than with the carbene. Balch also noted that,
although the position of the α-carbon atom in the vinyl
complex was not determined with certainty in the X-ray
experiment, the carbon atom bound to the pyrrolenine nitrogen atom is too far from the iron atom (2.83 Å) to be
considered bound to it and it is, therefore, most reasonable
to assume that there is a hydrogen atom between them.

There are several significant differences between the
structures of the N-methyl and N-vinylidene complexes. The
angle involving the substituted nitrogen atom, the iron atom
and the chloride ligand (N1-Fe-Cl) is 8° larger in the case
of the N-methyl complex and the opposite angle (N3-Fe-Cl) is
10° smaller. The tendency for the coordination geometry
about the iron atom to be less tetrahedral in the case of
the N-methyl complex is consistent with the trend found for
the zinc(II) complexes discussed above (i.e., the more reactive sp^3 hybridized methyl substituent produces a less tetrahedral complex than the sp^3 hybridized phenyl substituent). Another similarity with the trends for the zinc(II)
complexes is that the nitrogen atom bound to the sp^3 hybri-

dized substituent in the case of the iron(II) complexes, vinylidene is farther from the metal atom than is the nitrogen atom bound to the sp^3 hybridized methyl substituent (by 0.18 Å for the iron(II) complexes and by 0.04 Å for the N-phenyl and N-methyl complexes of zinc(II)). The cant of the substituted pyrrolenine ring, which is 44° for the vinylidene complex and 30° for the methyl complex of iron(II), is 42° and 39°, respectively, for the phenyl and methyl complexes of zinc(II). Evidently, the trends in structural differences between sp^3 and sp^3 substituents are in a similar direction but even more pronounced for these iron(II) complexes.

Table 2.9. Comparison of Structural Parameters of Some Iron Porphyrin Complexes.

Bond Lengths, Å	N-vinylidene[a]	N-methyl[b]	Carbene[c]
Fe-N1	2.510(12)	2.329(2)	2.299(1)
Fe-N2	2.133(11)	2.118(2)	1.991(4)
Fe-N3	2.087(12)	2.082(2)	2.002(4)
Fe-N4	2.131(11)	2.116(2)	1.985(4)
Bond Angles, deg			
N1-Fe-C1	95.5(3)	103.48(6)	120.5(1)[d]
N1-Fe-N2	80.6(4)	82.18(8)	82.8(2)[d]
N1-Fe-N3	140.3(4)	114.13(8)	127.8(2)[e]
N1-Fe-N4	81.0(4)	82.25(8)	83.1(2)[d]
N2-Fe-C1	102.3(3)	105.36(7)	97.0(1)
N2-Fe-N3	87.6(5)	87.12(8)	92.2(2)
N2-Fe-N4	143.8(4)	145.60(9)	164.5(2)
N3-Fe-C1	124.2(3)	114.38(7)	111.7(1)
N3-Fe-N4	86.9(4)	86.64(8)	91.4(2)

a Chloro-N-methyltetraphenylporphinatoiron(II), Ref. 3.
b Chloro-2,2-bis(p-chlorophenyl)-N-vinyltetraphenylporphi-natoiron(II), Ref. 20, note that the numbering is consis-tent with the convention in this book, not the original article.
c Chloro-N-bis(2,2-p-chlorophenyl)vinylidinotetrakis(p-tolyl)porphinato-1,N',N'',N'''-iron(III), Ref. 21.
d Here the carbon atom bound to the iron atom replaces N1.
e Non-bonded separation.

The final structure to be discussed in this section is a most unusual N-substituted metalloporphyrin in which two of the pyrroleninic nitrogen atoms are substituted and in which the two substituents form cycles with their terminal oxygen atoms bound to the iron atom, as shown in Figure 2.7.[22], [23] The structure of this complex is unique with respect to the structures of other N-substituted metalloporphyrins. This is the only reported case of a six-coordinate N-substituted metalloporphyrin. Binding of the two axial ligands at opposite sides of the porphyrin planes leads to a structure in which the iron atom is in-plane with respect to the four central nitrogen atoms. As a result, the iron-nitrogen bond distances for the alkylated nitrogen atoms are 2.242(5) $\mathring{A}$, in contrast to the 2.329(2) $\mathring{A}$ found for $Fe(N-CH_3TPP)Cl$. The distances between the iron atom and the unsubstituted nitrogen atoms in this structure are considerably shorter, 2.082 $\mathring{A}$, but this distance is still longer than the average for Fe(II) porphyrins, about 2.000 $\mathring{A}$. Balch and coworkers attributed this difference to the need for the core to expandto accommodate the lone pairs of electrons on the sp^3 hybridized substituted nitrogen atoms. Like the N-methyl and N-vinylidene complexes of Fe(II), this complex is high spin (S=2). Some of the important bond lengths and angles are given in Table 2.10. It should be noted that the trends among bond lengths and angles are similar to those for other N-substituted metalloporphyrins. The cant of the substituted pyrrolenine rings is especially pronounced as a result of the position of the in-plane iron atom.

This structure is not only interesting because it is novel, but also because it provides the first example of a structurally-characterized metallocyclic N-substituted porphyrin of the type that has been proposed as an intermediate in the formation of N-alkylated protoporphyrins from cytochrome P-450 enzymes. The role of such intermediates is discussed in Chapter 7.

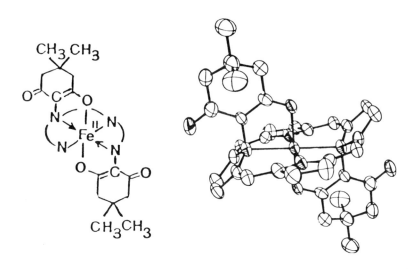

Figure 2.7 The structure of a bis-metallocyclic-iron(II)-bis-N,N''-dialkyl-5,10,15,20-tetraphenylporphyrin.[23]

Table 2.10 Selected Bond Lengths and Angles for a Bis-metallocyclic-iron(II)-N,N'-dialkyltetraphenylporphyrin.[a]

	Bond Lengths, Å		
Fe–N1, Fe–N3	2.242(5)	Fe–N2, Fe–N4	2.082(5)
Fe–O	1.998(6)	N1–C, N3–C	1.501(6)
$N1-C_a$, $N3-C_a$	1.444(4)	$N2-C_a$, $N4-C_a$	1.380(4)
C_m-C_a, alkylated	1.382(4)	C_m-C_a, nonalkylated	1.416(4)
C_a-C_b, alkylated	1.418(4)	C_a-C_b, nonalkylated	1.431(4)
C_b-C_b, alkylated	1.372(6)	C_b-C_b, nonalkylated	1.351(6)
	Bond Angles, degrees		
N1–Fe–N2	89.6(3)		
N1–Fe–O	84.1(2)		
N2–Fe–O	89.1(2)		
$N-C_a-C_b$, alkylated	108.0(3)	$N-C_a-C_b$, nonalkylated	109.6(3)
$C_a-C_b-C_b$, alkylated	109.1(3)	$C_a-C_b-C_b$, nonalkylated	107.3(3)
$N-C_a-C_m$, alkylated	127.0(1)	$N-C_a-C_m$, nonalkylated	124.6(3)
cant[b], alkylated	50.3	cant, nonalkylated	11.0

[a] These are average values for the two independent molecules in the asymmetric unit of the crystal. Estimated standard deviations are in parentheses. Ref. 23.

[b] The angles between the plane of the pyrroleninic ring and the plane of the iron atom and the four nitrogen atoms.

Structures of Bridged N-Substituted Complexes

Callot, Weiss, Balch and Mansuy and their coworkers coworkers have reported structures of three very unusual types
of metalloporphyrins: a type in which a carbene fragment
bridges from the porphyrin ring nitrogen atom to the metal
atom (N-ethoxycarbonylmethyleno-5,10,15,20-tetraphenylpor-
phinato-C,N',N'',N'''-nickel(II),[24] and chloro-N-bis(2,2'-p-
chlorophenyl)vinylidino-tetrakis(p-tolyl)porphinato-
1,N',N'',N'''-iron(III)[21, 25]); a second type are N-aminoporphyrin complexes in which the nitrogen atom of the amine
bridges a porphyrin ring nitrogen atom and the metal atom
(N-tosylamino-5,10,15,20-tetraphenylporphinato-N(amino)-
N',N'',N'''-nickel(II),[26] and chloro-N-(tosylaminochloromer-
cury(II))octaethylporphinato-N',N'',N'''-mercury(II)[27]); and
a third type in which two nitrogen atoms are bridged by a
methylene substituent and a palladium(II) atom bonds to the
other two ring nitrogen atoms (N,N'-((benzyloxy)methy-
lene)tetraphenylporphyrin-N'',N''')-dibromopalladium(II).[28]
The syntheses and reactions of these species are interesting
in their own right and will be discussed subsequently.
These and other structures are shown in Figure 2.8.
    The first of these unusual structures was reported in
1975 by Callot, Weiss and their coworkers[24, 29] and is of
interest with respect to the biological action of such
substances as DDT and, in addition, as models for oxidase
intermediates (*vide infra*). In the case of the N-CHCO$_2$C$_2$H$_5$
nickel(II) complex of tetraphenylporphyrin
(TPP(CHCO$_2$C$_2$H$_5$)Ni), the porphyrin ring is very nonplanar,
with angles between the individual pyrrole rings and the
best plane of the four porphyrin ring nitrogen atoms of
46.4^0 (N1), 13.3^0 (N2), 5.9^0 (N3) and 3.6^0 (N4).[30] Two
features of this structure that are in marked contrast with
the structures on N-alkyl and N-aryl complexes previously
discussed are the orientation of the substituted pyrrole
ring toward the metal atom rather than away from it and the
bonding of an atom of the N-substituent to the metal atom.

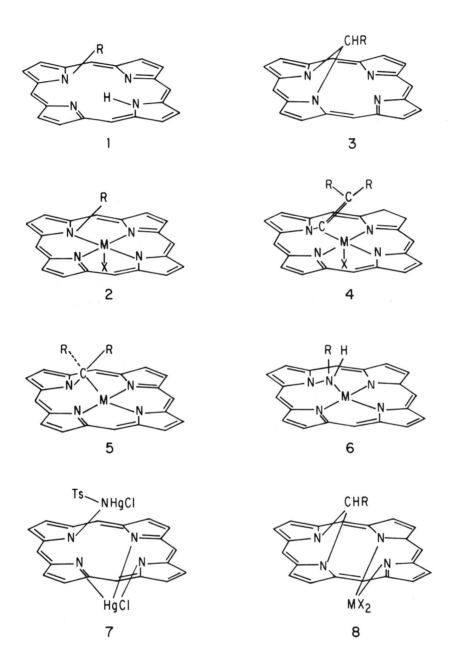

Figure 2.8. Structures of a variety of N-substituted porphyrins.

In the structure of the bridged nickel complex, the four nitrogen atoms are nearly coplanar (N1 and N3 deviate by − 0.04 Å and N2 and N4 by 0.04 Å), where the positive direction is towards the metal atom. These deviations are comparable in structures of N-alkyl porphyrin complexes (*e.g.*, deviations in Fe(N-CH$_3$TPP)Cl are 0.047 Å for N1 and N3 and −0.047 Å for N43). The methylene carbon is displaced from the nitrogen atom plane by 1.04 Å in the same direction as the Ni(II) atom (which is 0.19 Å from the plane). Although the porphyrin core is considerably distorted in this structure, the electron delocalization in the ring is not measureably affected (in contrast with N-alkyl-porphyrin complexes) with average values of N-C$_a$ = 1.382(6), C$_a$-C$_b$ = 1.434(8), C$_b$-C$_b$ = 1.344(6) and C$_a$-C$_m$ = 1.398(7) Å, values in the normal range for non-N-substituted metallo-porphyrins.[13,24] Structural parameters are summarized in Table 2.11.

Table 2.11. Selected Bond Lengths and Bond Angles in N-Ethoxycarbonylmethyleno-5,10,15,20-tetraphenyl-porphinato-C,N′,N′′,N′′′-nickel(II)[a]

		Bond Lengths, Å	
Ni–N2	1.911(3)	Ni–C	1.905(4)
Ni–N3	1.910(3)	N1–C	1.409(7)
Ni–N4	1.928(3)		
		Bond Angles, deg	
N2–Ni–N3	94.2(1)	C–Ni–N2	87.8(1)
N3–Ni–N4	93.8(1)	N2–Ni–N4	167.1(1)
N4–Ni–C	87.9(1)	N3–Ni–C	160.4(2)

a Ref. 24, note that atom number designations correspond to the system of this book, not the original article.

The structure of the vinylidene carbene complex chloro-N-bis(2,2'-p-chlorophenyl)vinylidino-5,10,15,20-tetrakis(p-tolyl)porphinato-1,N',N'',N'''-iron(III)[21] (TpTP((p-ClC$_6$H$_4$)$_2$C=C)FeCl) shows several significant differences from that of the nickel(II) carbene complex discussed above, in large part due to the higher coordination number of the Fe(III) atom. The iron atom is bound to three of the porphyrin nitrogen atoms (N2–N4), the carbene carbon atom (C(1)) and the chlorine atom. The coordination geometry is a distorted trigonal pyramid with N2 and N4 on the axis (with an N2–Fe–N4 angle of 164.5(2)$^\circ$) and C, N3, Cl and the iron atom forming the equatorial plane (with equatorial ligand–equatorial ligand and axial ligand–iron–equatorial ligand angles all within 8° of the ideal angles of 120° and 90°, respectively). The Fe–N bond lengths (Table 2.9) are within the normal range for Fe(III) porphyrin complexes and about 0.1 Å shorter than those in Fe(N–CH$_3$TPP)Cl, as expected for a comparison of Fe(III) and Fe(II) complexes. The Fe–Cl bond length of 2.299(1) Å is close to that in Fe(N-CH$_3$TPP)Cl (2.244(1) Å) and considerably different from that found in Fe(TPP)Cl (2.192(12) Å) or in chloro-protoporphyrin IX dimethyl ester iron(III) (2.218(6) Å). In this structure, the iron atom lies on the side of the porphyrin nitrogen plane toward the chlorine atom (0.31 Å) and the carbene carbon atom bound to the iron is on the opposite side (by 1.13 Å). The orientations of the pyrrole rings with respect to the metal atom are more similar to those in the N-alkyl-porphyrin complexes than to the Ni(II)-carbene complex: the N-substituted pyrrole is tilted away from the iron atom at an angle of 28° with respect to the plane of the four porphyrin nitrogen atoms while the other three pyrrole rings are tilted toward the iron atom at a much shallower angle. A structural aspect in which the Ni(II) and Fe(III) carbene complexes resemble each other and differ from the N-alkyl-porphyrin complexes is the apparent lack of perturbation of electron delocalization in the porphyrin ring system.[21] In these carbene complexes, the substituted porphyrin nitrogen atom retains its sp^3 hybridization.

Among the products obtained from the reaction of azides with porphyrins or zinc porphyrins, Callot and coworkers have isolated N-aminoporphyrins. These species form complexes and two have been analyzed crystallographically: the nickel(II) complex of N-tosylaminoTPP[29,30] and a bis(mercury(II)) complex of N-tosylaminoOEP.[27] The structure of the Ni(II) complex (Table 2.12) strongly resembles the structure of the Ni(II) N-carbene complex discussed above (Table 2.11). The four porphyrin nitrogen atoms are nearly coplanar (with an average absolute deviation of 0.07 Å), the Ni(II) atom is 0.21 Å from the plane toward the tosylamino nitrogen atom (N5) and N5 is 0.94 Å above the plane. The pyrrole rings are all rotated toward the tosylamino group, with dihedral angles with respect to the four-nitrogen plane of 40.5° (N1), 1.8° (N2), 4.4° (N3), and 9.4° (N4), each a few degrees less than the rotations in the carbene complex. The averaged values of bond lengths in the porphyrin skeleton are: $N-C_a = 1.382(7)$, $C_a-C_b = 1.437(9)$, $C_b-C_b = 1.340(8)$ and $C_a-C_m = 1.397(8)$ (where the value in parentheses is the greatest value of the estimated standard deviation for any of the individually determined bond lengths). These are not siginificantly different from those in the Ni(II) carbene complex and are very similar to those of non-N-substituted metalloporphyrins. In the N-tosylamino complex, the phenyl rings are rotated by angles of 73.0°, 76.8°, 86.1° and 51.9°, while in the Ni(II) carbene complex the corresponding values are 56.8°, 86.7°, 74.7° and 70.6°. For $Fe(N-CH_3TPP)Cl$ the angles have been reported with respect to the plane of the three non-N-substituted nitrogen atoms: 49.1°, 52.1°, -62.3° and -71.9°.[3]

Table 2.12. Selected Bond Lengths and Bond Angles in N-Tosylamino-5,10,15,20-tetraphenyl-N(amino),N',N'',N'''-nickel(II)[a]

Bond Lengths, Å

Ni-N2	1.920(3)	Ni-N5	1.830(4)
Ni-N3	1.883(4)	N1-N5	1.380(5)
Ni-N4	1.920(4)		

Bond Angles, degrees

N2-Ni-N3	94.0(1)	N5-Ni-N2	87.2(1)
N3-Ni-N4	94.1(1)	N2-Ni-N4	167.1(2)
N5-Ni-N4	87.3(1)		

----

a   Ref. 30.

The bis(chloro)mercury(II) complex of N-tosylamino-octa-ethylporphyrin has a much different structure, with one mercury atom nearly linearly coordinated to one nitrogen atom (of the N-substituted tosyl moiety) and a chlorine atom (bond angle = 162.1(2)°) and the other bound to three porphyrin nitrogen atoms and a chlorine atom (N2-Hg-Cl, 125.1(1)°; N3-Hg-Cl, 148.0(2)°; N4-Hg-Cl, 105.8(1)°; N2-Hg-N3, 79.3(2); N2-Hg-N4, 115.3(2)°, N3-Hg-N4, 75.4(2)).[27] The mercury-nitrogen bond distances are: Hg(1)-N2, 2.325(5) Å; Hg(1)-N3, 2.194(5) Å; Hg(1)-N), 2.496(5) Å and Hg(2)-N(amino), 2.075(5) Å and the mercury-chloro distances are: Hg(2)-Cl(2), 2.285(2) Å and Hg(1)-Cl(1), 2.318(2) Å. As found for the other N-tosylamino complex and the N-carbene complexes, the four porphyrin nitrogen atoms are nearly coplanar (within ± 0.03 Å). The N(amino) atom is further from the four-nitrogen plane in this complex (1.36 Å in the direction of the bound mercury atom, Hg(2), which is 2.05 Å from the N1-N4 plane than the Ni(II) N-tosylamino complex, however). The other mercury atom, Hg(1), is out-of-plane by 1.28 Å in the opposite direction. The angles between the pyrrole rings and the N1-N4 plane are: 40.8° (N1 ring, tilted toward Hg(2)), -10.2° (N2), -9.1° (N3) and -1.6° (N4 ring, *i.e.*, the latter three pyrrole rings are tilted away from Hg(2) and toward Hg(1)). The relative orientation of the pyrrole rings in this bimetallic complex follows the pattern of N-alkylporphyrin complexes rather than that of the Ni(II) N-tosylamino complex or the N-carbene complexes discussed above, but the bond distances in the porphyrin ring (N-C_a, 1.382(8) Å; C_a-C_b, 1.443(9) Å; C_b-C_b, 1.36(1) and C_a-C_m, 1.389(9) Å) are within the normal range of non-N-substituted metalloporphyrins. Thus, the factor which causes a change in π electron delocalization is the ability of the N-substituted nitrogen atom to bond (albeit less strongly than the other porphyrin nitrogen atoms) to the metal atom.

A unique structure among the N-substituted porphyrins has been reported by Callot, Fischer and Weiss for (N,N'-((ben-zyloxy)methylene)tetraphenylporphyrin-N'',N''')-dibromopal-ladium(II).[31] In this complex, the benzyloxymethylene group spans two nitrogen atoms (N1 and N2) while a palladium atom is bound to the other two porphyrin nitrogen atoms and to two bromine atoms. The methylene carbon atom is nearly tetrahedral (N-C-O angles of 109.1(6)° and 107.9(6)°) and is oriented on the opposite side of the N1-N4 plane from the palladium atom. The distance between this atom and the palladium atom is long - it represents a non-bonding distance. The distance from the methylene hydrogen atom to the palladium atom, however, is sufficiently short (2.58 Å) for them to be considered bound, giving a distorted square prism coordination sphere (the basal bond angles are: N-Pd-N, 79.0(2)°; Br-Pd-Br, 93.14(4)°; N-Pd-Br, 93.9(2)° and 93.8(2)°).[28] The two pyrrolenine rings bound to the methylene moiety are tilted toward the methylene carbon atom at angles of 8.2° and 12.7° with respect to the plane composed of the four meso carbon atoms and the two pyrrole rings bound to the palladium atom are tilted in the opposite sense (toward the Pd(II) atom) at angles of 17.4° and 21.3°. In this structure, the phenyl rings across from each other are nearly eclipsed and adjacent rings are of alternate rotational orientation. The dihedral angles between the phenyl rings and the meso-carbon plane are: 68.3° for the ring between the pyrroles bearing the methylene moiety (N1 and N2), 116.5° for the ring between the pyrroles bound to the Pd(II) atom (N3 and N4) and 68.0° and

120.8° for the others (between N1 and N4 and between N2 and N3, respectively). The porphyrin ring bond distances are within the normal range (N-C_a, 1.388(3) Å; C_a-C_b, 1.421(4) Å; C_b-C_b, 1.362(6) Å; C_a-C_m, 1.403(4) Å, where average estimated deviations are given in parentheses).

The relatively normal bond lengths in the porphyrin ring system of the complexes discussed in this chapter are consistent with the general similarities that have been found when visible absorption and nmr spectra of these N-substituted complexes are compared with those of non-N-substituted porphyrins. The spectral properties of N-substituted porphyrins and their metal complexes will be examined in detail in the following chapter. The structural similarities in the aromatic system, despite the marked lack of planarity due to substitution at the nitrogen atom, demonstrate the remarkable ability of the porphyrin ring system to accommodate a variety of bonding patterns. It would be interesting to investigate the thermodynamic properties of these compounds (such as their heats of atomization and/or heats of hydrogenation) to determine the actual extent of aromaticity retained when the nitrogen atom is substituted. At present, the degree to which aromaticity is conserved can only be inferred in a qualitatitive sense from the structural results presented in this chapter and the spectroscopic results presented in the following chapter.

References

1. D.K. Lavallee and O.P. Anderson, J. Amer. Chem. Soc., **104** (1982) 4707-4708.

2. O.P. Anderson and D.K. Lavallee, Inorg. Chem., **16** (1977) 1634-1640.

3. O.P. Anderson, A.B. Kopelove and D.K. Lavallee, Inorg. Chem., **19** (1980) 2101-2107.

4. O.P. Anderson and D.K. Lavallee, J. Amer. Chem. Soc., **98** (1976) 4670-4671.

5. O.P. Anderson and D.K. Lavallee, J. Amer. Chem. Soc.,
   99 (1977) 1404-1409.

6. D.K. Lavallee, O.P. Anderson and A. Kopelove, J. Am.
   Chem. Soc., 100 (1978) 3025-3033.

7. C.K. Schauer, O.P. Anderson, D.K. Lavallee, J.P. Battioni
   and D. Mansuy, J. Amer. Chem. Soc., 109 (1987)

8. D. Kuila, D.K. Lavallee, C.K. Sauer and O.P. Anderson,
   J. Am. Chem. Soc., 106 (1984) 448-450.

9. G.M. McLaughlin, J. Chem. Soc., Perkin II, (1974) 136-140.

10. D.E. Goldberg and K.M. Thomas, J. Am. Chem. Soc.,98
    (1976) 913-917.

11. A.M. Abeysekera, R. Grigg, J. Trocha-Grimshaw and
    K. Hendrick, Tetrahedron, 36 (1980) 1857-1868.

12. E.B. Fleischer, Accts. Chem. Res., 3 (1970) 105-112.

13. J.L. Hoard, in Porphyrins and Metalloporphyrins, K.M.
    Smith, ed., Elsevier, Amsterdam, 1975, p. 317 ff.

14. L.J. Radonovitch, private communication.

15. G.B. Jameson, F.S. Molinaro, J.A. Ibers, J.P. Collman,
    J.I. Brauman, E. Rose and K.S. Suslick, J. Am. Chem.
    Soc., 100 (1978) 6769-6770.

16. J.P. Collman, J.L. Hoard, G. Lang, L.J. Radonovitch,
    C.A. Reed, J. Am. Chem. Soc., 97 (1975) 2676-2680.

17. D.K. Lavallee and D. Kuila, Inorg. Chem., 23 (1984) 3987-
    3992.

18. J.-P. Battioni, D. Lexa, D. Mansuy and J.-M. Savéant,
    J. Amer. Chem. Soc., 105 (1983) 207-215.

19. C.E. Castro and R.S. Wade, J. Org. Chem., 50 (1985)
    5342-5851.

20. A.L. Balch, Y.W. Chan, M.M. Olmstead and M.W. Renner, J.
    Org. Chem., 51 (1986) 4651-4656.

21. M.M. Olmstead, R.-J. Cheng and A.L. Balch, Inorg. Chem.,
    21 (1982) 4143-4148.

22. D. Mansuy, P. Battioni, J.-F. Bartoli and J.-P. Mahy,
    Biochem. Pharmacol., 34 (1984) 431-432.

23. J.-P. Battioni, I. Artaud, D. Dupré, P. LeDuc, I. Akhrem,
    D. Mansuy, J. Fisch, R. Weiss and I. Morgenstern-
    Badarau, J. Amer. Chem. Soc., 108 (1986) 5598-5607.

24. B. Chevrier and R. Weiss, _J. Am. Chem. Soc._, **98** (1976) 2985–2990.

25. B. Chevrier, R. Weiss, M. Lange, J.-C. Chottard and D. Mansuy, _J. Am. Chem. Soc._, **103** (1981) 2899–2901.

26. H.J. Callot and E. Schaeffer, _J. Organomet. Chem._, **145** (1978) 91–99.

27. H.J. Callot, B. Chevrier and R. Weiss, _J. Am. Chem. Soc._, **101** (1979) 7729–7730.

28. H.J. Callot, J. Fisher and R. Weiss, _J. Am. Chem. Soc._, **104** (1982) 1272–1276.

29. H.J. Callot, Th. Tschamber, B. Chevrier and R. Weiss, _Angew. Chem._, **87** (1975) 545.

30. H.J. Callot, B. Chevrier and R. Weiss, _J. Am. Chem. Soc._, **100** (1978) 4733–4741.

CHAPTER 3

CHEMISTRY OF N-SUBSTITUTED PORPHYRINS II: SPECTROSCOPIC PROPERTIES

Visible-UV Absorption Spectroscopy

Certainly one of the most useful physical properties of a porphyrin is its intense visible-uv absorption spectrum, which typically allows characterization of micromolar solutions. The N-substituted porphyrins are similar to unsubstituted porphyrins in this regard, with molar absorptivities in the Soret region (400-500 nm) of 10^5 and 10^4 in the of 500-700 nm range. The similarity in molar absorptivities to those of non-N-substituted porphyrins indicates that the high degree of aromaticity typical of nearly planar porphyrins is retained in the N-substituted porphyrins even though they are highly distorted.

The absorption bands of N-substituted porphyrins (Tables 3.1 and 3.2 and Figures 3.1 and 3.2) are generally red-shifted on the order of 10 nm from bands of corresponding non-N-substituted porphyrins. Spectra of the free bases typically have four pronounced bands in the visible region (500-700 nm). Compared with spectra of the free bases, spectra of monoprotonated and diprotonated N-substituted porphyrins generally show comparable molar absorptivities but fewer absorption maxima. The marked difference in the cant of the substituted pyrrolenine ring, which was discussed in the previous chapter, does not allow the dication to assume the inversion center symmetry found for dications of non-N-substituted porphyrins such as H_4TPP^{2+}, however. Disubstituted or bridged porphyrins in which the neighboring nitrogen atoms (designated N,N' or N_a, N_b) are substituted, often have simplified spectra like the dications (Tables 3.3 and 3.4). The spectra of porphyrins N-substituted on opposite rings are like those of mono-N-substituted porphyrins.

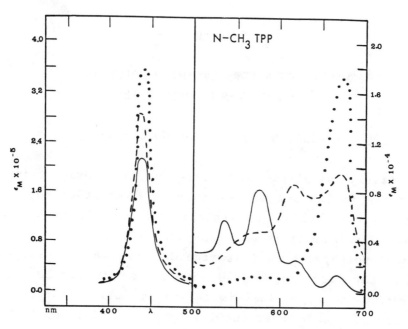

Figure 3.1. Visible absorption spectra of N-methyl-5,10,15,20-tetraphenylporphine (solid line), the monocation (dashed line) and the dication (dotted line) in DMF.

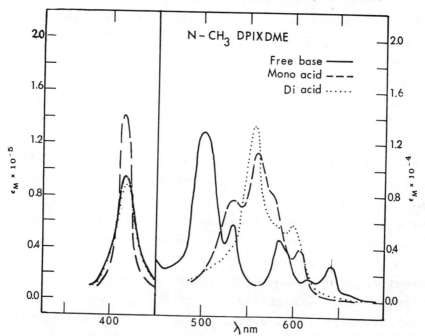

Figure 3.2. Visible absorption spectra of N-methyldeuteroporphyrin IX dimethyl ester, monocation and dication.[2]

Table 3.1. Visible Absorption Spectra of Some N-Substituted
meso-Tetraphenylporphyrins

Porphyrin	Solvent	Maxima (Log of Absorptivity)	Ref
N-CH$_3$TPPHTPP	CH$_2$Cl$_2$	432(5.35), 534(3.99), 575(4.18) 613(3.67), 676(3.70)	1
N-CH$_2$CH$_3$HTPP	CH$_2$Cl$_2$	432, 531, 573, 613, 675	3
N-CH$_2$CO$_2$C$_2$H$_5$- HTPP	CH$_2$Cl$_2$	437(5.58), 532(4.13), 574(4.30) 617(3.81), 678(3.84)	4
N-CH$_2$C$_6$H$_4$HTPP	CH$_2$Cl$_2$	434(5.56), 533(4.12), 573(4.30) 615(3.85), 675(3.82)	5
N-CH$_2$C$_6$H$_4$NO$_2$- HTPP	CHCl$_3$	432(5.46), 527(4.04), 568(4.20) 614(3.65), 676(3.70)	6
N-CH(CH$_3$)CO$_2$CH$_3$- HTPP	C$_6$H$_6$	435(5.44), 532(3.99), 575(4.15) 622(3.62), 683(3.71)	7
N-C$_6$H$_5$HTPP	CHCl$_3$	444(5.59), 550(4.24), 575(4.56) 633(4.14), 705(3.94)	8
N-C$_6$H$_5$H$_3$TPP^{2+}	CHCl$_3$(TFA)	457(5.71), 694(4.94)	8
N-CH$_3$H$_2$TPP$^+$	CHCl$_3$(TFA)	442(5.44), 626(4.32), 677(4.36)	9
N-CH$_3$H$_3$TPP^{2+}	CHCl$_3$(TFA)	445(5.54), 678(4.68)	9
N-C(H)C(C$_6$H$_4$Cl)$_2$- HTPP		435(5.34), 529(4.08), 569(4.18) 622(3.61), 684(3.70)	10
N-CH$_2$C$_6$H$_5$- HTCPP^{4-}	CH$_2$Cl$_2$	430(5.18), 536(4.01), 577(4.23) 676(3.79)	11
N-CH$_2$C$_6$H$_5$- H$_2$TCPP^{3-}	CH$_2$Cl$_2$	433(5.25), 574(4.02), 613(4.01) 669(4.01)	11
N-CH$_3$HTPPS$_4$$^{4-}$	H$_2$O	435(5.43), 540(3.94), 584(4.20) 610(3.97), 672(3.78)	11
N-CHC(p-ClC$_6$H$_4$)$_2$$^-$ HTPP		435(4.34), 497(sh), 592(4.08) 569(4.18), 622(sh), 684(3.70)	10

Table 3.2. Visible Spectra of Some Non-*meso*-substituted N-Alkylporphyrins

Porphyrin	Solvent	Maxima (Log of Absorptivity)	Ref
N–CH_3HOEP	$CHCl_3$	412, 508, 539, 586, 616, 642	12
N–$CH_2CO_2C_2H_5$–HOEP	CH_2Cl_2	451(5.11), 556(3.95), 606 (4.00) 666(3.70)	13
N–$CH_2CO_2C_2H_5$–HOEP	CH_2Cl_2	408(5.17), 505(4.13), 540(3.81) 578(3.72), 631(3.51)	13
N–$CH_3H_3OEP^{2+}$	$CHCl_3$	405, 551, 596	14
N–CH_3HPP		422, 514, 548, 595, 652	15
N–$CH_3H_3PP^{2+}$	$CHCl_3$	412, 560, 604	16
N–CH_2CH_2OHHPP	$CHCl_3$	419, 510, 546, 594, 650	17
N–$C_2H_3H_3PP^{2+}$	$CHCl_3$	418, 563, 606	18
N–$CH_2CH_2CH_3$HPP	$CHCl_3$	422, 514, 549, 593, 650	19
N–$CH_2CH_2CH_3H_3PP^{2+}$	$CHCl_3$	418, 565, 610	19
N–$CH_2C_6H_5$HPP	CH_2Cl_2	417(5.04), 511(4.04), 544(3.83) 627(3.34), 651(3.34)	20
N–CH_3HDP	$C_6H_5NO_2$	405(5.19), 501(4.15), 531(3.77) 583(3.67), 614(3.19), 641(3.41)	21
N–$CH_3H_2DP^+$	$C_6H_5NO_2$	413(5.15), 535(3.87) 560(4.06) 578(sh, 5.17), 606(3.58), 641(3.45)	21
N–$CH_3H_3DP^{2+}$	$C_6H_5NO_2$	416(4.93), 556(4.13), 599(3.76)	21
N–CH_3HMP	$CHCl_3$	412, 506, 537, 586, 615, 642	19
N–$CH_2CH_2CH_3$HMP	$CHCl_3$	413, 507, 539, 587, 613, 643	19
N–$CH_3H_3MP^{2+}$	$CHCl_3$	406, 552, 596	19
N–$CH_2CH_2CH_3H_3MP^{2+}$	$CHCl_3$	412, 559, 600	19
N–CH_3HEtio		503, 533, 587, 645	22

Table 3.3. Visible Absorption Spectra of Several Di- and Tri-N-substituted Porphyrins

Porphyrin	Solvent Maxima (Log of Absorptivity)	Ref
$N-CH_3,N'-CH_3TPP$	$CHCl_3$ 460(5.26), 642(4.19), 710(4.02)	9
$N-CH_3,N'-CH_3-HTPP^+$	$CHCl_3$ 442(5.38), 624(4.33), 669(4.33) (w/TFA)	9
$N-CH_3,N'-CH_3-H_3TPP^{2+}$ (*trans*)	$CHCl_3$ (w/TFA) 453(5.44), 693(4.68)	9
$N-CH_3,N'-CH_3,N''-CH_3TPP^+$	$CHCl_3$ 465(5.04), 684(4.30), 714(4.32)	9
$N-CH_3,N'-CH_3,N''-CH_3HTPP^{2+}$	$CHCl_3$ (w/TFA) 467(5.39),724(4.72)	9
$N-CH_3,N'-CH_3,N''-CH_2CH_3TPP^+$	$CHCl_3$ 466(5.14), 686(4.29), 715(4.34)	9
$N-CH_3,N'-CH_3,N''-CH_2CH_3HTPP^{2+}$	$CHCl_3$ (w/TFA) 468(5.43), 728(4.69)	9
$N-CH_3,N'CH_3-H_2OEP^{2+}$ (*trans*)	$CHCl_3$ (w/TFA) 413(5.32), 564(3.16), 609(3.67)	23
$N-CH_3,N''-CH_3-H_2OEP^{2+}$ (*trans*)	$CHCl_3$ (w/TFA) 400(5.28), 538(3.85), 564(3.95) 572(3.88), 613(3.26)	23
$N-CH_3,N''-CH_3-H_2OEP^{2+}$ (*cis*)	TFA 412(5.39), 568(3.99), 620(3.68)	23
$N-CH_3,N'-CH_3,N''CH_3$ OEP$^+$ (*trans, trans*)	TFA 418(5.23), 536(2.97), 560(3.61) 574(3.76), 604(3.52), 624(3.43)	23

Table 3.4. Visible Absorption Spectra of Several
N,N'-Bridged Porphyrins.

Porphyrin	Solvent	Maxima (Log of Absorptivity)	Ref
N, N'– C(H)CH$_3$TPP	CH$_3$C$_6$H$_5$	432(5.26), 532(4.20), 608(3.94)	6
N, N'– C(H)CH$_2$CH$_3$TPP	CH$_3$C$_6$H$_5$	434(5.27), 533(4.19), 608(3.96)	6
N,N'– C(H)CH(CH$_3$)$_2$TPP	CH$_3$C$_6$H$_5$	435(5.18), 533(4.13), 608(3.91)	6
N, N'– C(H)(CH$_2$)$_2$OCH$_3$TPP	CH$_3$C$_6$H$_5$	435(5.13), 533(4.08), 608(3.89)	6
N, N'– CH$_2$C$_6$H$_5$TPP	CH$_3$C$_6$H$_5$	437(5.13), 535(4.10), 610(3.86)	6
N, N'– N–tosylaminoTPP	CH$_2$Cl$_2$	432(5.34), 510(3.36), 548(3.97) 585(4.00), 640(3.90)	24
N, N'– N–p–nitrobenzylaminoTPP	CH$_2$Cl$_2$	427(5.27), 505(3.50), 544(4.04) 583(4.02), 640(3.89)	24
N, N'– C(H)OCH$_3$OEP	CH$_3$C$_6$H$_5$	415(4.85), 518(4.02), 545(3.77) 588(3.83), 640(3.69)	6
N, N'– C(C$_6$H$_4$Cl)$_2$T(p–OCH$_3$P)P	CH$_2$Cl$_2$	453(5.02), 560(3.82), 605(4.29) 645(3.90)	25
N, N'– C(C$_6$H$_4$Cl)$_2$HTPP	C$_6$H$_6$	430(5.15), 509(3.89), 549(4.08) 585(4.19), 631(3.85)	10

a In all cases, the bridges are bound to neighboring
   pyrroleninic nitrogen atoms in a *cis* (same side) fashion.

Metal complexes of N-substituted porphyrins (Tables 3.5 and 3.6 and Figures 3.3 and 3.4) typically show three strong bands in the region of 500–700 nm rather than the two bands which are commonly found for non-N-substituted metallo-porphyrins.[2,26] The Zn(II) and Cd(II) complexes of N-alkyl and N-phenyl tetra-*meso*-substituted porphyrins have partially resolved split Soret bands and complexes of non-*meso*-substituted porphyrins have widely-separated double Soret bands. Although it is reasonable to assume that these variations from the normal pattern are due to the lower symmetry of the N-substituted complexes (approximately mirror plane symmetry at the coordination site rather than four-fold rotation symmetry), there have not been any theoretical studies to corroborate this assumption.

As is evident from the data in Tables 3.1 and 3.2, very similar spectra obtain for a variety of N-substituted por-phyrins in which the functional group attached to the por-phyrin nitrogen atom is a methylene moiety. The spectra of N-phenyl porphyrins and their complexes are shifted signifi-cantly from those of the corresponding N-methylene species. From the comparison of the molecular structures of the chlorozinc(II) complexes of N-methyl-5,10,15,20-tetraphenyl-porphyrin and N-phenyl-5,10,15,20-tetraphenylporphyrin,[3] the spectral differences appear to be due to electronic differences in the N-substituents rather than structural differences.

One of the most interesting differences in the spectra of the N-substituted porphyrins and corresponding non-N-substi-tuted porphyrins is found when a series of complexes of different metal ions is compared. For a particular N-substituted porphyrin ligand, these spectra are much more similar to one another (Tables 3.5 and 3.6). It should be noted, of course, that nearly all the complexes of N-substi-tuted porphyrins which have been spectroscopically charac-terized are divalent and that spectral shifts would be expected for differences in the valence of the metal atom. Ogoshi has reported the spectra for Fe(III) complexes of N-

methyloctaethylporphyrin which have Soret bands more than 30
nm to the blue relative to Co(II) and Zn(II) complexes of N–
CH$_3$OEP.[27], [28], [29] The great similarity in spectra of
divalent metal atoms has been discussed specifically for
complexes of 5,10,15,20–tetraphenylporphyrin,[26] N–methyl–
deuteroporphyrin dimethyl ester[2] and N–methylprotoporphyrin
dimethyl ester,[30] and is a general characteristic of all N–
alkyl and N–phenyl porphyrins reported to date. One known
exception to this similarity occurs in of 5,10,15,20–tetra–
phenylporphinatocopper(II). These spectra show a marked
dependence on the presence of either strongly coordinating
ligands (such as Cl$^-$ or amines) or with very weakly
coordinating counterions (such as CF$_3$SO$_3^-$ or ClO$_4^-$).[31]

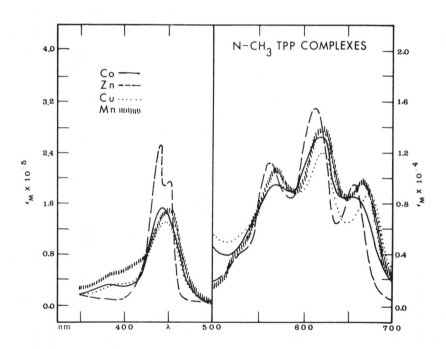

Figure 3.3. Visible absorption spectra of several divalent
metal complexes of N–methyl–5,10,15,20–tetraphenylporphine
(with chloride as the axial ligand) in dichloromethane.

Table 3.5. Visible Spectra of Metal Complexes of N-Substituted
Meso-Tetraphenylporphyrins

Complex	Solvent	Max.(Log of Molar Absorptivity)	Ref
Co(N-CH$_3$TPP)Cl	CH$_2$Cl$_2$	360(4.52), 445(5.19), 575(3.98) 624(4.12), 661(3.93)	2
Co(N-C$_2$H$_5$TPP)Cl	CH$_2$Cl$_2$	438, 454(sh), 564, 610, 659	32
Cu(N-CH$_3$TPP)Cl	CH$_2$Cl$_2$	385(sh,4.51), 448(4.98) 565(3.98),627(4.08), 671(3.95)	26
Cu(N-CH$_3$TPP)Cl	CH$_3$CN	371, 451, 564, 617, 672	31
Cu(N-C$_2$H$_5$TPP)Cl	CH$_3$CN	371, 450, 568, 616, 671	33
Cu(N-PhTPP)Cl	CH$_3$CN	378, 459, 569, 627, 689	33
Cu(N-CH$_3$TPP)$^+$	CH$_2$Cl$_2$	430, 442, 546, 597, 656	34
Cu(N-C$_2$H$_5$TPP)$^+$	CH$_2$Cl$_2$	434(5.04), 441(5.08), 547(3.95) 590(3.85), 657(4.08)	34
Cu(N-PhTPP)$^+$	CH$_2$Cl$_2$	440(5.04), 455(5.11), 556(4.01) 609(3.99), 675(3.74)	34
Cu(N-CH$_2$C$_6$H$_5$NO$_2$)$^+$	CH$_2$Cl$_2$	441(5.24), 547(3.14), 599(3.01) 659(2.86)	34
Fe(N-CH$_3$TPP)Cl	THF	447, 459, 564, 610, 662	3
Fe(N-CH$_3$TPP)Br$^+$	CHCl$_3$	415, 450(sh), 530, 565, 620, 675	35
Fe(N-C$_2$H$_5$TPP)Cl	THF	446, 457, 563, 612, 662	3

Continued

Table 3.5. Continued.

$Fe(N-PhTPP)Cl$	CH_2Cl_2	454(4.96), 466(sh), 569(3.85) 630(4.00), 682(3.85)	36
$Fe(N-PhTPP)^+$	PhCN	453(4.95), 465(sh), 568(3.78) 631(3.95), 682(3.78)	36
$Fe(N-PhTPP)^{2+}$	PhCN	375(sh), 432(4.75), 570(sh) 662(3.78), 697(3.78)	36
$Mn(N-CH_3TPP)Cl$	CH_2Cl_2	368(4.97), 452(5.19), 567(4.04) 624(4.15), 666(3.99)	26
$Mn(N-C_6H_5NO_2)Cl$	CH_3CN	450, 540, 567, 621, 666	5
$Mn(N-PhTPP)Cl$	CH_3CN	457, 468, 572, 638, 689	5
$Ni(CH_2CO_2C_2H_5-$ HTPP$)^+$	CH_2Cl_2	451(5.11), 556(3.95) 606(4.00), 666(3.70)	13
$Zn(N-CH_3TPP)Cl$	CH_2Cl_2	439(5.41), 448(5.28), 561(4.07) 611(4.19), 658(3.98)	26
$Zn(N-PhTPP)Cl$	CH_2Cl_2	447, 459, 567, 638, 680	26
$Zn(N-CH_2CO_2CH_3-$	CH_2Cl_2	442(5.26), 564(3.92) 615(4.19), 660(3.93)	37
$Zn(N-CH_2COO-$ $C_2H_5TPP)^+$	CH_2Cl_2	442(5.26), 564(4.03) 615(4.20), 660(4.01)	37
$Zn(N-CH(CH_3)-$ $COOCH_3TPP)^+$	CH_2Cl_2	444(5.13), 564(3.90) 620(4.04), 675(3.90)	37
$Zn(N-CH-$ $(COOCH_3)_2TPP)^+$	CH_2Cl_2	444(5.21), 562(4.16) 620(4.35), 670(4.18)	37

Table 3.6. Visible Spectra of Metal Complexes of
N-Substituted Porphyrins Which Are Not *meso*-Substituted

Porphyrin	Solvent	Maxima (Log of Absorptivity)	Ref
Co(N-CH$_3$OEP)OAc	CHCl$_3$	379(sh), 428(4.80), 542(3.77) 588(3.92), 628(sh)	28
Co(N-CH$_3$DP)Cl	CH$_2$Cl$_2$	384(sh), 424(4.78), 538(3.90) 583(3.78)	2
Fe(N-C$_6$H$_5$OEP)$^+$	PhCN	395(sh), 434(4.96), 450(sh) 544(4.04), 597(4.04)	36
Fe(N-C$_6$H$_5$OEP)$^{2+}$	PhCN	383(4.91), 545(3.95), 604(4.00)	36
Fe(N-C$_6$H$_5$OEP)Cl$^+$	CHCl$_3$	388(4.97), 504(3.85) 566(3.70), 661(sh)	27
Mn(N-CH$_3$DP)Cl	CH$_2$Cl$_2$	384(4.74), 427(4.82), 534(3.91) 585(3.97)	2
Rh$_2$Cl$_2$(CO)$_4$-OEP	CHCl$_3$	398(4.56), 430(4.48), 555(3.70) 595(3.75), 617(3.54)	38
Zn(N-CH$_3$OEP)$^+$	CHCl$_3$	422, 538, 585, 625	14
Zn(N-CH$_3$DP)Cl	CH$_2$Cl$_2$	419(5.13), 429(5.06), 534(3.90) 572(sh), 583(4.08), 623(3.25)	2
Zn(N-CH$_3$PP)$^+$	CHCl$_3$	431, 547, 596, 634	14
Zn(N-C$_2$H$_5$PP)$^+$	CHCl$_3$	431, 545, 592, 632	18
Zn(N-CH$_2$CH$_2$OH)-PP)Cl	CHCl$_3$	423, 546, 591, 633	17

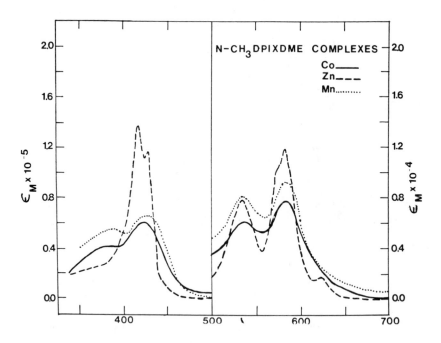

Figure 3.4. Visible absorption spectra of several divalent metal complexes of N-methyldeuteroporphyrin IX dimethyl ester (with chloride as axial ligand) in dichloromethane.

The general similarity of the uv–visible absorption spectra of N-substituted porphyrin complexes relative to the spectra of non-N-substituted porphyrin complexes may arise directly from two features of their molecular structure. The most obvious characteristic of the crystallographically-determined structures is the domed shape of the coordination site, in which the metal ion lies significantly out of the porphyrin plane (see Table 2.2). In this structure, the $d\pi$–$p\pi$ overlap would not be as great as in the planar structures found for many non-N-substituted porphyrins. In addition, for all of the complexes of N-substituted porphyrins whose spectra are available, the metals are high spin. This spin state is consistent with the limited coordination number (four strong bonds – three to the unsubstituted nitrogen atoms and the fourth to the axial ligand) and the non-planar conformation of the porphyrin coordination site, which provides a weaker ligand field than that of a typical planar metalloporphyrin. With a weaker ligand field and less $d\pi$-$p\pi$ interaction, it would be expected that the variation in d orbital occupancy would not cause as great a change in uv-visible spectra. A second important consideration is that there is simply much less variety among reported structures of N-substituted complexes than there is for the non-N-substituted porphyrin complexes. Monometallic non-N-substituted metalloporphyrin complexes can be four, five or six coordinate with planar, square-base pyramidal, octahedral or irregular coordination geometries and can be high spin, low spin or of intermediate spin, leading to quite different electronic spectra.

A very useful aspect of the visible absorption spectra of the metal complexes of N-substituted porphyrins is related to their behavior in weakly acidic solution. As will be discussed further in Chapter 4, the metal atom in an N-substituted porphyrin complex is typically displaced by acid much more readily than is a metal atom bound in a non-N-substituted porphyrin complex. Therefore, treatment of a solution that is suspected to contain an N-substituted metalloporphyrin with millimolar acid will typically lead to a change in the 500-700 nm region from a spectrum typical of the N-substituted metalloporphyrin (typically three bands) to a spectrum typical of a diprotonated porphyrin cation and rapid neutralization of that solution will lead to the spectrum of the free base form of the porphyrin. Similar treatment of many non-N-substituted metalloporphyrins will lead to very little change in the visible spectrum.

Fluorescence Spectroscopy

Very few fluorescence spectra have been reported for N-substituted porphyrins and their complexes.[2,39] Those fluorescence spectra which are available have general features similar to those of non-N-substituted porphyrins. The quantum yields of the free-base porphyrins are on the order of 0.1, with the overall luminescence (due to the combination of absorbance, efficiency of radiationless transition to the emitting singlet state vs. intersystem crossing and relative stability of the emitting excited state to deactivation via radiationless transition) at a given concentration generally somewhat greater for the N-substituted porphyrin. Just as the absorption spectra show red shifts of N-substituted porphyrins with respect to corresponding unsubstituted porphyrins, the excitation spectra of N-substituted porphyrins also show red shifts. The emission spectra of the N-substituted porphyrins are red-shifted as well, but they show the pattern typical of corresponding unsubstituted porphyrins; two prominent emission maxima likely corresponding to two vibrational manifolds of the emitting excited state.

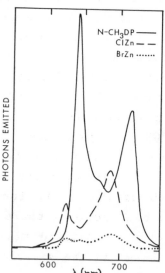

Figure 3.5. Fluorescence spectra of N-methyldeuteroporphyrin IX dimethyl ester (solid line), its chlorozinc(II) complex (dashed line) and its bromozinc(II) complex (dotted line), in CH_2Cl_2 at $20°C$ using 410 nm excitation.

The quantum yields for Zn(II) complexes are of the same order as those of free base porphyrins, but depend on the nature of the axial ligand: the greater the number of electrons, the lower the luminescence intensity. For instance, the quantum yield for Zn(N-CH$_3$DP IX DME)Br is about one-half that of Zn(N-CH$_3$DP IX DME)Cl, as shown in Figure 3.5. The quantum yield for diamagnetic metal atoms also follows the normal heavy-atom effect, decreasing in the series: Zn(II), Cd(II), Hg(II). N-substituted porphyrin complexes of paramagnetic atoms such as high spin Mn(II), Fe(II) and Co(II) show no fluorescence at the detection limit of a quantum yield of 10^{-5}. Some non-N-substituted metalloporphyrins show little or no fluorescence at room temperature but exhibit relatively strong phosphorescence at low temperatures (for example, Cu(II) porphyrins[40,41]). No phosphorescence data for N-substituted porphyrins are available for comparison.

For porphyrins, there are two advantages to luminescence spectrometry relative to absorption spectroscopy: greater sensitivity and greater selectivity. With typical commercial fluorimeters, it is feasible to determine porphyrin (and Zn(II) porphyrin) concentrations to 10^{-9}M while the limit for the typical visible absorption spectrometer is about 10^{-6}M. Those porphyrins which fluoresce (free bases and diamagnetic complexes) can be distinguished from non-fluorescent species in a particular sample and porphyrins can often be distinquished from one another by differences in both their excitation and emission spectra. The red shifts of the N-substituted porphyrins thus allow them to be distinguished fluorimetrically in the presence of the corresponding non-N-substituted compounds. Another very useful feature of fluorescence spectroscopy is that the technique readily distinguishes complexes, in which a metal ion is covalently bound, from ion pairs, in which the metal ion is present in a counterion. It is also possible by this technique to monitor either acid dissociation of a metal complex or a reaction sequence in which a diamagnetic ion is replaced by a paramagnetic ion.

Vibrational Spectroscopy

Infrared spectra of a number of N-CH$_3$TPP complexes and Raman spectra of two metal-nitrogen bridged tetraphenylporphyrin have been reported.[42], [43] The infrared spectra were obtained in the region of 200-1600 cm^{-1} for chloro-N-methyltetraphenylporphinato complexes of Mn(II), Fe(II), Co(II), Ni(II), Cu(II), Zn(II) and Cd(II) as well as the phenylmercury(II) complex of N-CH$_3$TPP. In order to make assignments, spectra were also obtained of Co(N-CH$_3$TPP)Br and isotopically pure complexes [64]Zn(N-CH$_3$TPP)Cl and [68]Zn(N-CH$_3$TPP)Cl and the deuterated species, Co(N-CD$_3$TPP)Cl. The former complexes allow unambiguous assignment of the metal-halide stretching bands; and the deuterated complex aids in the assignment of bands with considerable contribution from N-C vibrations of the substituted nitrogen atom.[42]

With respect to the general structural properties of N-substituted porphyrins, two regions of the infrared spectrum are of considerable interest: the so-called metal sensitive bands near 1000 cm^{-1} and the metal halide stretching bands near 300 cm^{-1}. Boucher and Katz reported variations of bands from 920-970 cm^{-1} for protoporphyrin IX and hematoporphyrin IX dimethyl ester complexes.[44], [45] Similarly, Thomas and Martell found bands near 1000 cm^{-1} that shift to progressively higher energies for a series of complexes of tetraphenylporphyrin in the same order as Buchler's stability series.[46], [47] The shift in this band in the series from CdTPP to PtTPP is 996 to 1019 cm^{-1}. In the case of the chloro-N-methylTPP complexes, a strong band is found at 1005{2 cm^{-1} for Mn(II), Co(II), Ni(II), Cu(II), Zn(II), Cd(II) and phenylmercury(II) and at 1002 cm^{-1} for Fe(II).[42] However, distinct differences exist in this region for the N-CH$_3$TPP complexes. Two other strong bands, one at 999$\pm$1 cm^{-1} and another at 986$\pm$1 cm^{-1} differ for each complex relative to the band near 1005 cm^{-1} (Figure 3.6). Based on

the normal coordinate analysis of Ogoshi[48] and the fact that
deuteration of the $N-CH_3$ group does not affect these bands,
they appear to be due chiefly to peripheral vibrational
modes (such as C-H rock involving the pyroleninic hydrogen
atoms).   From   the   crystal   structures   of   the $N-CH_3$TPP
complexes discussed in this chapter, it is apparent that the
C-C bonds in the N-substituted pyrrolenine ring are
different from those in the other pyrrolenine rings.   Since
the bond between the metal atom and the substituted nitrogen
atom differs in length for each metal atom, the pyrrolenine
ring could be affected sufficiently to alter the vibrational
spectrum.

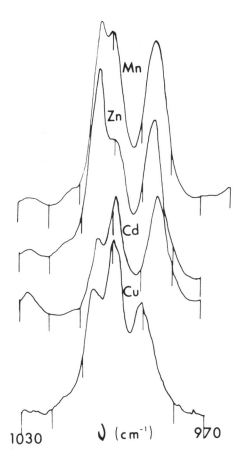

Figure 3.6.   Infrared absorption spectra of N-methyl-
5,10,15,20-tetraphenylporphyrin complexes in the metal-
sensitive region.

The Resonance Raman spectra of the metal-nitrogen bridged complexes Fe,N-[C=C(p-ClC$_6$H$_4$)$_2$](TPP)Cl and Ni,N-(CHCO$_2$C$_2$H$_5$)-(TPP) were compared with the reference compounds Fe(TPP)Cl and NiTPP to determine the effect of reduced symmetry.[43] Three differences were apparent: 1) new vibrational modes were evident for the bridged species in the region below 1000 cm^{-1}, 2) the two highest porphyrinic vibrational modes (near 1550 cm^{-1}) were lower by 13 to 31 cm^{-1} for the bridged compounds, and 3) the depolarization ratios are more highly dispersed, as expected for a reduction from D$_{4h}$ symmetry to C$_s$ symmetry.[43]

Nuclear Magnetic Resonance Spectroscopy

One of the first spectroscopic properties reported for an N-substituted porphyrin was the remarkable upfield nmr chemical shift of the N-methyl protons in N-methyloctaethylporphyrin.[49] The tilt of the N-substituent from the plane of the porphyrin nitrogen atoms places the protons of the N-substituent out of the plane of the porphyrin "ring current". As a result, the protons on the carbon atom bound to the pyrrolic nitrogen are shifted upfield relative to TMS rather than downfield. For N-methyl protons of a wide variety of porphyrins, the resonance occurs at about 4 ppm upfield (Table 3.7 - 3.12 and Figures 3.7 and 3.8). The effect of the ring current is attenuated by distance as illustrated by the nmr chemical shifts for the protons of ethyl and substituted ethyl groups and for N-phenyl and N-benzyl substituents. Since the N-substituted porphyrins are relatively strong bases and can exist as a mixture of free base, monocationic and dicationic species in the presence of even trace amounts of water, it is often more convenient to use the nmr spectra of their Zn(II) complexes, which are diamagnetic, rather than those of the metal-free porphyrin. The N-substituents also exhibit chemical shifts several ppm upfield from normal positions (Tables 3.13-3.15). The region of the proton nmr spectrum above TMS is so generally clear of interfering resonances that this characteristic feature of the N-substituted porphyrins may make their detection possible even in complex mixtures. The major drawback to this technique, of course, is its relative insensitivity.

Proton Magnetic Resonance Spectra of Uncomplexed Synthetic Porphyrins. One of the interesting features of the proton NMR spectra of N-substituted porphyrins is the difference among the four pyrrolenine rings caused by N-substitution. In many cases, the proton resonances of the four rings can be clearly distinguished. For synthetic porphyrins such as the *meso*-tetraarylporphyrins, the two pyrroleninic protons of the N-substituted ring are magnetically equivalent, as are the two of the opposite ring (due to the reflection plane passing through N1 and N3). In each of the adjacent rings, there are two inequivalent protons — one nearer the N-substituted ring and the other farther from it. The protons of the two rings either side of the N-substituted ring (which contain N2 and N4) are related by the reflection plane, giving rise to two doublets, each with an integrated intensity of two protons and the same coupling constant, as is clear in Figure 3.7. The tentative assignment for the pyrroleninic resonances (Table 3.7) is that the least downfield singlet (the most disturbed from its normal position by N-substitution) is due to the protons of the N-substituted pyrrolenine ring and, therefore, the other singlet is due to the protons of the opposite ring. The two doublets cannot be definitely assigned without further data (such as that obtainable from Nuclear Overhauser Enhancement experiments).

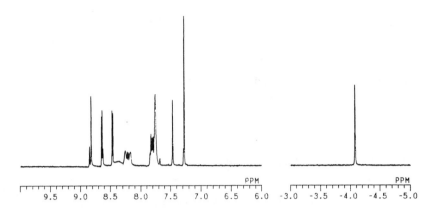

Figure 3.7. The ^{1}H NMR spectrum of N-methyl-5,10,15,20-tetraphenylporphyrin in CDCl$_3$, 10^{-2} M, with TMS.

Table 3.7. ^{1}H NMR Shifts of N-Substituted Porphyrins: Mono-N-Alkyltetraphenylporphyrins

Compound (Solvent)	N-CH$_x$	N-H	pyrrolenine				phenyl		Ref
			s	d	d	s	o,m	m,p	
other protons									
N-CH$_3$HTPP (CDCl$_3$)	-4.10		7.46	8.46	8.64	8.82	8.1-8.4	7.8	9
N-CH$_3$H$_2$TPP$^+$ (CDCl$_3$)	-4.80		7.67	8.61	8.73	8.42	8.3-8.7	8.0	9
N-CH$_3$H$_3$TPP^{2+} (TFA)	-3.97	-4.08(1H) -3.97(2H)	7.83	8.96	9.15	8.91	8.5-8.9	9.2	9
N-C$_2$H$_5$HTPP (CDCl$_3$)	-4.48	-1.72(-CH$_3$)	7.50	8.46	8.66	8.78	8.1-8.4	7.76	32

Table 3.7. Continued

Compound (Solvent) N-CH$_x$ N-H other protons	Shifts (Downfield from TMS) pyrrolenine				phenyl		Ref
	s	d	d	s	o,m	m,p	

N-CH$_2$C$_6$H$_4$NO$_2$HTPP
(CDCl$_3$)   -3.38   7.66 8.56 8.72 8.96   8.0-8.3   7.8   6
            ca. 0(1H)
   4.67(2H, N-sub. phenyl,o), 7.48(2H, m)

N-C$_6$H$_5$HTPP   50
(C$_6$D$_6$)   7.35 8.16 8.33 8.70
   2.99(2H, N-sub. phenyl,o), 5.20(2H, m), 5.67(1H,p)

N-C$_6$H$_5$NO$_2$HTPP   24
(CDCl$_3$)   8.11 8.92 9.16 8.89   7.7-8.35
              0.6(2H)
   3.44(2H, nitrophenyl,o), 6.36(2H, m)

N-CH$_2$CO$_2$C$_2$H$_5$HTPP   32
(CDCl$_3$)   8.02 8.44 8.47 8.67   8.1-8.34 7.44
   -1.22(3H, ester), 0.88(2H, ester)

N-C(CH$_3$)(H)CO$_2$CH$_3$HTPP   7
(C$_6$D$_6$)   -3.73   8.44 8.46 8.65 8.75   7.2-8.5
              -2.51(3H)
   2.27(3H, ester)

N-C(CH$_3$)(H)CO$_2$C$_2$H$_5$HTPP   7
(C$_6$H$_6$)   -3.84   8.44 8.46 8.65 8.75   7.47-8.5
   -2.54(3H)
   0.09(3H, ester), 2.66(2H, ester)

N-tosylaminoTPP   24
(CDCl3)   -0.05 (N-H) 7.85 8.86(br)   8.7   7.6-8.5

N-CHC(p-ClC$_6$H$_4$)$_2$HTPP   10
(CDCl$_3$)   7.51 8.39(AB,4H) 8.22   7.64-8.14

Some of the trends in NMR spectra of tetraphenylporphyrins are evident in Table 3.7. The spectra of simple N-alkyl derivatives (N-methyl-, N-ethyl-, N-propyl- and others such as N-2-hydroxyethylporphyrins) are very similar and only two entries are given in the table. From the crystal structure of a monocationic species discussed in Chapter 2, and from marked alteration in the visible absorption spectrum (Figure 3.1), protonation of N-CH$_3$HTPP to give N-CH$_2$H$_2$TPP$^+$ and N-CH$_3$H$_3$TPP^{2+} can be expected, to lead to significant changes in the angles which the various pyrrolenine rings make with one another. These changes are, indeed, evident changes in the NMR spectra. There appears to be a regular downfield progression of the shifts of the *meso*-phenyl protons, the singlet assigned to the protons of the N-substituted ring, and the two doublets assigned to the neighboring rings, but little about structural changes can be directly deduced from such data. The shifts are small enough that it is difficult to determine the relative amounts of free base or cationic species actually present unless a good integration can be obtained for the N-H resonance, which is often quite broad. For this reason, many N-substituted porphyrins have been characterized by the NMR spectra of their zinc(II) complexes, avoiding the possibility of equilibria between free base and cationic forms. These are discussed in the next section.

Marked differences in the NMR spectra of free base N-substituted tetraphenylporphyrin derivatives are found for a variety of N-aryl, N-acyl and other substituents other than simple alkyl groups. The relatively similar crystal structures that we have found for N-benzyl, N-methyl and N-phenyl derivatives (discussed in Chapter 2) and the similarity in their visible absorption spectra leads us to suspect that the changes in NMR chemical shifts caused by these substituents are more likely due to electronic than structural effects. In this regard, the pronounced differences in shifts in the proton NMR spectra of N-phenylHTPP and N-p-nitrobenzylHTPP are interesting. In the case of the N-ethylcarbonylethylene derivative, the resonance assigned to the protons of the N-substituted ring differs from that of N-CH$_3$HTPP, tempting speculation that the equilibrium position of the carbonyl group may be close to this pyrrolenine ring. Again, N.O.E. experiments would be interesting.

The resonances of the aryl protons of N-substituents vary considerably. The greatest change from the normal phenyl resonances occurs for N-phenylHTPP, the next for N-p-nitro-benzylHTPP and somewhat less for N-benzylHTPP (Table 3.7). Significantly less effect is evident for N-vinylporphyrins with a phenyl ring bound to the carbon that is attached to the pyrrolenine nitrogen atom. (Table 3.8). Perhaps the phenyl ring in the vinylidene derivatives is located under the center of the porphyrin ring rather than under the pyrrolenine ring, as in the structure of the zinc(II) complex of N-benzylHTPP.[51]

Table 3.8. ^{1}H NMR Shifts of N-Vinyltetraphenylporphyrins[a]

Compound	Shifts (Upfield from TMS)				
	N-CRC(H)-	N-C(CH)$_x$	phenyl substituents		
			ortho	meta	para
N-C(C$_6$H$_5$)CH$_2$HTPP	-1.81	1.18(H)	5.55	6.85	ca 7.1
N-C(C$_6$H$_5$)CHCH$_3$HTPP	-1.36	-1.36(CH$_3$)	5.61	7.1	7.1
N-C(C$_6$H$_4$NO$_2$)CHNO$_2$HTPP	-1.70	1.22(H)	5.58		

a Taken in CDCl$_3$, Ref. 52.

It is expected that the resonance for the protons on the carbon atom attached to the pyrroleninic nitrogen would depend on the nature of that carbon atom and on the state of protonation of the other pyrroleninic nitrogens. In all cases of metal-free species, these resonances lie significantly upfield from TMS. For mono-N-alkylporphyrins (Tables 3.7-3.9) these are typically in the range of -3.4 to -5.3 ppm (upfield) relative to TMS. For multiply alkylated derivatives (Tables 3.10 and 3.11), the resonances occur at even higher field.

Table 3.9. ^{1}H NMR Shifts of Non-*Meso*-Substituted N-Substituted Porphyrins

Compound (Etio type)	Shifts (Downfield from TMS)									Ref
N-CH_x / *meso*-H	ring CH_3				ring CH_3 (of $-CH_2CH_3$) ring CH_2 (of $-CH_2CH_3$)					
	1	3	5	7	2	4,8	6			
N-CH_3HEtio (CDCl$_3$)	3.21	3.50	3.52	3.60	1.46	1.84(4,6,8)				53
N-CH_x = −4.76; *meso*-H = 9.86, 9.88					3.71	3.98	4.11			
N-CH_3H$_2$Etio$^+$	3.17	3.58	3.64	3.64	1.24	1.80	1.84	1.66		53
N-CH_x = −5.13; *meso*-H = 10.49, 10.63						3.7–4.2				
N-CH_3H$_3$Etio^{2+} (TFA)	3.36	3.83	3.83	3.83	1.47	1.94	1.90			53
N-CH_x = −5.29; *meso*-H = 10.90, 11.00; −4.21 (N–H)					3.90	4.35	4.35			

(OEP and OMP type)	N–CH$_x$ / *meso*-H	N–H	1,2 ring CH$_3$ ring CH$_2$	3,4,7,8	5,6	Ref
N-CH_3HOMP	−4.76 / 9.86, 9.94		3.13	3.44	3.56	23
N-CH_3H$_2$OMP$^+$ (TFA)	−4.23 / 11.16	−4.23	3.33	3.79	3.82	23
N-CH_3HOEP (CDCl$_3$)	−4.76 / 9.89, 9.94		1.48 / 3.72	1.90 / 3.96	1.91 / 4.00	23
N-CH_3H$_2$OEP$^+$ (CDCl$_3$)	−5.18 / 10.55, 10.64		1.44	1.90 / 3.85–4.84	1.91	23
N-CH_3H$_3$OEP^{2+} (TFA)	−5.20 / 11.00, 11.12	−3.96	1.54 / 3.84	2.00 / 4.40	1.94 / 4.40	23
N-$CH_2CO_2C_2H_5$HOEP (CDCl$_3$)	−4.27 / 9.88, 10.00; 0.22(ester CH$_3$)	−3.55	(1.43 1.83 1.86 1.90)[a] / (3.6–4.3)			13

[a] Not assigned specifically.

Table 3.10. ^{1}H Shifts of Di- and Tri-methyltetraphenyl-porphyrins[a]

Compound (Solvent)	N-CH₃ N-H	pyrrolenine s	d	d	s	phenyl o,m	m,p
N,N'-(CH₃)₂-TPP[b]	-4.90	8.18	7.7	8.0	8.18	8.3-8.7	7.8
N,N'-(CH₃)₂-HTPP⁺ [b]	-4.77	(7.5-7.8)	7.71	7.99		8.2-8.8	7.9
N,N'-(CH₃)₃-H₂TPP²⁺ [d]	-4.42	1.12[c]	8.17	7.99	9.15[c]	8.5-8.9	8.2
N,N',N''-(CH₃)₃TPP⁺ [d]	-5.57(3H) -3.12(6H)	8.19	7.27	7.45	8.40	8.4-8.7	7.9
N,N',N''-(CH₃)₃HTPP²⁺[b]	-4.72(3H) -3.18(6H), -3.6(N-H)	7.3	7.78	7.2	8.9	8.6-8.9	8.2
N,N'-(CH₃)₂-N''-C₂H₅TPP⁺ [b]	-5.60(3H) -3.38(6H) -3.50 and -4.18(N-CH₂-), -0.62(N-CH₂CH₃-)	7.27	7.25	7.48	8.45	8.4-8.6	8.0
N,N'-(CH₃)₂-N''-C₂H₅HTPP⁺ [d]	-4.73(3H) -3.38(6H), -3.65(N-H) -3.5 and -4.15(N-CH₂-), -0.42(N-CH₂CH₃-)	8.32	7.80	7.20	9.19	8.5-8.9	8.2

Shifts (Downfield From TMS)

a Ref. 9, The di-methylporphyrins in this table are all in the *trans* configuration., b In CDCl₃, c These are doublets. d In deuterated trifluoroacetic acid.

Table 3.11.　　　^{1}H NMR Shifts of Di- and Tri-alkylated Octaethylporphyrins

Compound (Solvent)	N–CH$_3$ N–H meso-H	ring CH$_3$ ring CH$_2$ ring CH$_2$	Ref
N,N'-(CH$_3$)$_2$OEP (*trans*) (CDCl$_3$)	−5.92 −5.88 −3.76 10.21, 10.33	3.7–4.2 1.50(2), 1.70(4) 1.93(6), 1.91(8)	53
N,N'-(CH$_3$)$_2$OEP (*cis*) (CDCl$_3$)	−3.52 10.32	3.39–3.98 1.28(1,2,5,6) 1.84(3,4,7,8)	53
N,N''-(CH$_3$)$_2$OEP (*trans*) (CDCl$_3$)	−5.30 9.80	3.8–4.2 1.71(1,2,5,6) 1.94(3,4,7,8)	53
N,N',N''-(CH$_3$)$_3$OEP$^+$ (*trans,trans*) (CDCl$_3$)	−7.08(N') −3.92(N,N'') 10.04	3.3–4.1 1.63(1,6), 1.50(2,5) 1.73(3,4), 1.89(7,8)	53
N,N''-(CH$_3$)$_2$H$_2$OEP^{2+} (*cis*) (CDCl$_3$)	−3.70 −5.92 10.24	3.88–4.31 1.37–2.16	23
N,N''-(CH$_3$)$_2$H$_2$OEP^{2+} (*trans*) (TFA)	−5.11 11.40	3.3–4.1 2.20–1.97	23
N,N',N''-(CH$_3$)$_3$HOEP^{2+} (*trans,trans*) (TFA)	−6.28, −4.16 10.82, 10.90	3.8–4.5 1.4–2.2	23

Callot has reported the NMR spectra of a number of *cis*-bridged tetraphenylporphyrins synthesized by a remarkable phase-transfer reaction involving chloroform, base and an alcohol which produces the bridged porphyrin bearing the alkyl group from the alcohol to the carbon atom originating from the chloroform molecule.[6] An important characteristic of the NMR spectra of these compounds is the two pairs of doubled doublets arising from the pyroleninic protons (Table 3.12). For these species, there is a mirror plane (or, in the case of some unsymmetric substituents on the bridging carbon atom, possibly a near-mirror plane) which contains the bridging carbon atom, the hydrogen atom bound to it, and both the *meso*-carbon atom between the bridged nitrogen atoms, and the two opposed atoms (Figure 3.8 shows this structure as well as several others). Thus, the two pyrolenine rings bearing the bridging carbon atom each have one proton closer and one farther from the bridge; a similar symmetry holds for the other two pyroleninic rings, giving rise to two pairs of doublets with different coupling constants. The difference in symmetry between these complexes and the mono-N-substituted porphyrins is clearly and consistently evident in their NMR spectra. It is interesting to note that the hydrogen atom bound to the bridging carbon atom in these compounds is not free to move (as in the case of N-methylporphyrins, for example) and only a single sharp resonance is found in all cases. This result indicates that there is only one isomer. No crystal structures or N.O.E. data are available for these species, but the relatively small upfield displacement of the bridging hydrogen atom (about −1 ppm instead of the typical −3 to −4 ppm) seems more consistent with a position closer to the center of the porphyrin ring, as Callot has indicated in the figures in his articles (as in Figure 2.8).

Proton Magnetic Resonance Spectra of Zinc(II) Complexes of Synthetic Porphyrins. The diamagnetic zinc(II) complexes of N-substituted porphyrins are rapidly and quantitatively formed in organic solvents using zinc acetate or other zinc salts. In the absence of acid or chelating ligands, the complexes are typically stable (except for certain N-substituents which very readily stabilize carbocations; see Chapter 4). Since the binding of a metal atom to the coordination site prevents formation of cationic species by protonation, this method simplifies interpretation of NMR spectra, an especially important consideration for non-*meso*-substituted porphyrins, which tend to be highly basic. A few representative ^{1}H NMR parameters are given in Table 3.13. Interestingly, the shifts are quite comparable between the N-methyl protons of N-CH$_3$HTPP and Zn(N-CH$_3$TPP)Cl; N-CH$_3$Etio I and Zn(N-CH$_3$Etio)Cl; and N-CH$_3$HOEP and Zn(N-CH$_3$OEP)Cl. Yet, there is a significant difference between the ortho proton resonances of N-PhHTPP and Zn(N-PhTPP)Cl (0.6 ppm less upfield). In the last case, perhaps inductive effects as well as the through-space effect of the zinc(II) atom exist. Substantial differences in shifts occur for the pyroleninic resonances of the TPP complexes and free bases (1.3-1.5 ppm downfield shifts for the substituted ring upon complexation). The effect on the pyrroleninic methyl and methylene groups of etioporphyrin I and octa-ethylporphyrin are also substantial, although the greater distance of these protons from the aromatic porphyrin ring apparently diminishes the difference in shifts. Again, the most pronounced effect occurs for the protons of the substituted pyrrolenine ring and magnitude of the shifts for these two porphyrins is quite substantial.

Table 3.12. ^{1}H NMR Shifts of N,N'-Bridged Tetraphenylporphyrins[a]

Compound	N-C$\underline{H}$	N-COC$\underline{H}$	N-COCC$\underline{H}$	pyrrolenine[b] phenyl or *meso* H
N,N'-CHOCH$_3$-TPP	−1.18	−1.87		8.35,8.67(5.1);8.96,9.22(4.3) 7.8, 8.3
N,N'-CHOC$_2$H$_5$-TPP	−1.22	ca.−2.0		8.32,8.62(4.4);8.90,9.14(5.0) 7.8, 8.3
N,N'-CHOCH-(CH$_3$)$_2$TPP	−1.30	−2.92	−2.15	8.26,8.58(4.2);8.85,9.11(5.2) 7.8, 8.3
N,N'-CHOCH$_2$-OCH$_3$TPP	−1.20 (t, J =5.5 Hz)	−2.02	1.6	8.33,8.64(4.4);8.90,9.17(5.1) 7.8, 8.3
N,N'-CHOCH$_2$-C$_6$H$_5$TPP	−0.93 m and p of C$_6$H$_5$ at 6.50	−1.10	4.43	8.34,8.67(4.1);8.94,9.19(5.0) 7.8, 8.3
N,N'-CHOCH$_3$-OEP	−2.5	−2.5		3.7-4.6(−C$\underline{H}_2$−),1.92(−CH$_2$C$\underline{H}_3$) 9.96(1H), 10.00(1H), 10.42(2H)

a In CDCl$_3$, relative to TMS., Ref. 6.
b All are doublets. The table entries are pairs of doublets with the same coupling constant (the coupling constant is in parentheses).

Table 3.13. ^{1}H NMR Shifts of Some Zinc Complexes of
N-Substituted Porphyrins

Ligands	Shifts (^{1}H, Downfield from TMS)		Ref
	N-R　　　meso	pyrrolenine	
N-CH$_3$-Etio, Cl$^-$	-4.63　10.20, 10.25(2H)　　10.27	-CH$_3$　　　　　　　　3.52(1),3.57(3),3.62(5,7)　-C$\underline{H}$CH$_3$　　　　1.75(2),1.88(4,6,8),3.52(1)	53
N-CH$_3$OEP, Cl$^-$	-4.61　10.22, 10.31	-C$\underline{H}_2$CH$_3$　　3.98(1,2),4.04(3,8),4.08(4-7)　-CH$_2$C$\underline{H}_3$　1.75(1,2),1.94(3-8)	23
N-C$_6$H$_5$OEP, Cl$^-$	2.0(o),4.92(m)　　5.46(p)		54
N-CH=C(p-C$_6$H$_4$Cl)$_2$, Cl$^-$	-2.24	7.90(s),8.82(d),8.75(d),8.93(s)	55
N-CH$_3$TPP, Cl$^-$	-3.83　8.62,8.13,8.32(o),7.7-7.9(m,p)	8.30(s),8.84(d),8.97(d),8.92(s)	55
N-C$_6$H$_5$TPP, Cl$^-$	2.36(o),5.22(m)　　5.76(p)	8.92(s),8.74(d),8.66(d),8.48(s)	3
N-TosylaminoTPP	2.23(-CH$_3$), 5.63,6.73(2d,J=8Hz)　　7.41(s),8.89(d),9.00(d),8.83(s)		24

Table 3.13.   Continued

Ligands	N-CH$_x$	-CO$_2$CH$_x$	pyrrolenine 2,3	7,8,17,18[b]	12,13	Ref
N-CH$_2$CO$_2$CH$_3$HTPPBr$_4$						4
	-3.88	3.30	8.30	8.95	8.90	
N-CH$_2$CO$_2$C$_2$H$_5$HTPPBr$_4$						4
	-3.96	2.78	8.24	8.86,9.00	8.86	
N-CH$_2$CO$_2$CH$_3$HTPP, Cl$^-$						37
	-3.88	2.76	8.25	8.87,9.00	8.85	
N-CH$_2$CO$_2$C$_2$H$_5$HTPP, Cl$^-$						37
	-3.88	0.50	8.28	8.90,9.05	8.87	
N-CH(CH$_3$)CO$_2$CH$_3$HTPP, Cl$^-$						37
	2.3	2.23	8.28	8.87,9.04	8.80	
N-CH(CH$_3$)CO$_2$CH$_3$HTPP, Cl$^-$						37
		1.63	8.13	8.75,8.94	8.30	

a   In CDCl$_3$. The *meso* phenyl ring   proton resonances   occur
between 7.8 and 9.3 ppm for TPP complex. They   appear   as an
AA'BB multiplet   centered at   8.2 ppm   for the TPP Br$_4$ complexes.
b   AB   doublets,   J in the range of 4.0 to 4.3 Hz.

**Proton Magnetic Resonance Spectra of Complexes of Natur-
ally-Derived Porphyrins.** For the symmetric synthetic porphyrins such as tetraaryl-, octaethyl- and isomer I of etioporphyrin, the spectra are well-resolved because there is
only one geometric isomer. Note that for etioporphyrin I,
however, enantiomers exist. Synthetic alkylation of
nonsymmetric porphyrins such as protoporphyrin IX or its
dimethyl ester, however, typically produces a mixture
containing 25% of each of the four geometric isomers (and a
racemic mixture of the enantiomers of each). In many cases,
only the complex spectra of such mixtures are available.
However, in a few cases the mixtures have been separated by
high performance liquid chromatography (HPLC). The landmark
paper of this type, which led the way for discovering the
stereoselective alkylation of porphyrins *in vivo*, was
published in 1981 by Ortiz de Montellano and Kunze.[56] By
the careful use of N.O.E. data, they were able to assign the
spectra of the zinc(II) complexes of the four separated
isomers (Figure 3.9) As shown in Table 3.13, the shifts are
very similar for the four isomers but they can be readily
distinguished using typical high field NMR spectrometers.
For isomers I and III or isomers II and IV, the pyrroleninic
methyl resonances (3 vs. 5 and 1 vs. 8, respectively) or
the *meso*-proton resonances (gamma position in both cases)
provide larger differences than those of the N-methyl group
itself. Use of the isomeric identification of N-substituted
porphyrins will be discussed further in Chapter 7, the
biochemistry of N-substituted porphyrins arising from
cytochrome P-450 enzymes.

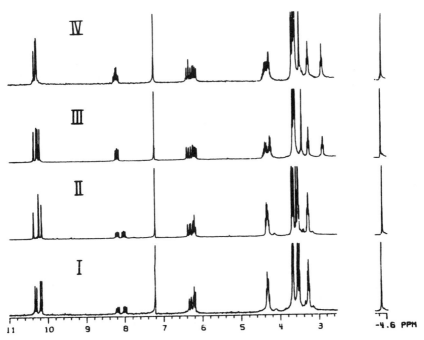

Figure 3.9. The ^{1}H NMR spectrum of the chloro-N-methylpro-toporphyrin IX dimethyl ester zinc(II) isomers[56] (the isomers have the methyl group on the ring as indicated: I, B; II, A; III, C; IV, D. The Roman numerals indicate the order of elution from an HPLC column.) The methyl proton signals were truncated. Each was in $CDCl_3$ with TMS.

Table 3.14. ^{1}H NMR Shifts for the Four Isomers of the Zinc Complexes of N-Methylprotoporphyrin IX Dimethyl Ester[a]

Group	I	Pyrrole Ring II	III	IV
N–Me	−4.492	−4.530	−4.497	−4.535
1–Me	3.659	3.548	3.630	3.663
3–Me	3.517	3.646	3.650	3.620
5–Me	3.554	3.526	3.429	3.566
8–Me	3.546	3.582	3.587	3.469
6–MeO	3.686	3.682	3.592	3.619
7–MeO	3.686	3.693	3.613	3.548
α–*meso*	10.338	10.373	10.344	10.315
β–*meso*	10.291	10.233	10.272	10.265
γ–*meso*	10.142	10.150	10.232	10.247
δ–*meso*	10.193	10.236	10.193	10.245

a Ref. 56, assignments based on NOE's.

A very important series of proton NMR data obtained by
Ortiz de Montellano and coworkers is given in Table 3.15.
These data represent the products obtained from treating
rats with the terminal alkenes ethylene, propene and octene
and the terminal alkyne octyne.[57] The green pigments formed
in the liver as a result of the interaction of these compounds with cytochrome P-450 enzymes were isolated and purified by HPLC. From the absorptivities of the visible absorption bands of the initial extracts and the purified
products, they estimated that the initial materials
contained only one major isomer (at least 90% of the total).
Thus, although the synthetic alkylation of protoporphyrin IX
produces a random distribution of isomers, the biological
reaction is highly stereoselective (not altogether expected
since the cytochrome P-450 enzymes are able to bind a wide
range of substrates, some of which, like the steroids, are
large). By using selective spin decoupling and differences
in the T_1 relaxation times of various proton resonances,
they firmly established the assignment of the isomers. The
T_1 relaxation times (not shown in Table 3.15) are shortest
for the methyl groups bound directly to the pyrrolenine
rings (positions 1, 3, 5 and 8) and longest for the $O-CH_3$
groups. Among the *meso* protons, T_1 for the gamma proton is
much shorter for all cases except the product of the propene
interaction (N-(2-hydroxypropyl)protoporphyrin IX). Spin
decoupling patterns are indicated by entries in Table 3.14.
The nature of the product was established from mass
differences using $^{16}O_2$ and $^{18}O_2$ to determine that there is a
hydroxyl group present and coupling patterns to deduce that
it is on the second carbon atom. In each case, the product
is obtained by first attaching the terminal atom of the
double or triple bond to the pyrroleninic nitrogen atom (to
the D ring for alkenes and the A ring for octyne) and then
adding a hydroxyl group to the other carbon atom of the
unsaturated bond. The mechanism of these reactions is
discussed in detail in Chapter 7.

Table 3.15. Signal Assignments for the Dimethyl Esterified Chlorozinc N-Alkylprotoporphyrin IX Derivatives[a]

Group	Porphyrin			
Product: N-2-hydroxy- ethane[b]		propane[b]	octane[b]	octene[c]
Administered: ethylene		propane	octene	octyne
N-CH$_x$	-4.9	-5.16,-4.95	-5.13,-4.94	-4.37
N-CH$_2$-CH$_x$[d]	0.63	0.9	0.76	
Methyl				
1	3.65,γ	3.66,γ	3.65,γ	3.42,γ
3	3.61,α	3.61,α	3.62,α	3.61,α
5	3.55,β	3.55,β	3.55,β	3.53,β
8	3.38,γ	3.35,γ	3.36,γ	3.53,γ
O-CH$_3$	3.70	3.70	3.69	3.70
O-CH$_3$	3.55	3.57	3.62[d]	3.69
Meso				
α	10.27	10.31	10.27	10.33
β	10.23	10.22	10.22	10.20
γ	10.32	10.28	10.26	10.15
δ	10.29	10.28	10.31	10.21
Internal Vinyl				
2	8.22,α, Me$_1$[e]	8.04	8.22, Me$_1$[e]	7.92,α
4	8.17,β, Me$_3$[e]	7.19	8.16, Me$_3$[e]	8.12,β
Sidechain Methylenes				
7	4.35	4.32	4.32	4.35
8	4.05, 4.25	4.02, 4.25	4.04, 4.21	4.35
Sidechain Methylenes				
7	3.23	3.22	3.21	3.31
8	3.08	3.03	3.02	3.31

a   Ref. 57 and 58.

b   Isomers with the N-substituent on ring D (bearing a propionic acid substituent).

c   Isomer with the N-substituent on ring A (bearing a vinyl substituent).

d   Other N-Alkyl assignments. Propene: CH$_3$, -1.2 ppm; Octene: 3-CH$_2$, -1.45 and -1.05 4-CH$_2$, -0.42 and -0.15; 5-CH$_2$, 0.15; 6-CH$_2$, 0.5; 7-CH$_2$, 0.7; and 8-CH$_3$, 0.5 ; Octyne: 3-CH$_2$, 0.08 ; 4-CH$_2$, -0.35; 5-CH$_2$, 0.08; 6-CH$_2$, 0.58; 7-CH$_2$, 0.75; and 8-CH$_3$; 0.5.

It is perhaps a bit surprising that the N-CH$_x$ protons of
N-(2-hydroxypropyl)- and N-(2-hydroxyoctyl)protoporphyrin IX
dimethyl ester zinc(II) appear as separate peaks (doublets)
rather than as a single peak. It appears that rotation
about the N-C bond is not rapid on the NMR time scale for
these species even though it is for the N-(2-hydroxyethyl)
derivative. We have observed a similar phenomenon (a pair
of doublets upfield 3.4 ppm from TMS) for N-benzylHTPP and a
complex series of doublets for the random mixture of isomers
of N-benzylprotoporphyrin IX (Lavallee, Battioni and Mansuy,
unpublished results).

The careful and highly credible analysis of the products
of the interactions of alkenes and octyne with cytochrome P-
450 represents an extremely significant contribution that
has been important in deducing a rational mechanism for
these reactions and it has even provided a means of determining the stereochemistry of the active site of cytochrome P-
450 enzymes (discussed in detail in Chapter 7).

**Proton Magnetic Resonance Spectra of Other Diamagnetic
Metal Complexes: Bridged Complexes.** Callot and coworkers
have presented NMR spectra of two types of N,N'-bridged
complexes: those with a carbon atom bridging the neighboring
pyrrolenine nitrogen atoms (N,N'-bridged porphyrins) and
those in which the bridging atom is bound to one pyrrolenine
nitrogen atom and to the metal atom (M,N-bridged porphyrins,
see Figure 3.9).[6,24]   Some representative examples of
diamagnetic bridged complexes of palladium(II), mercury(II)
and nickel(II) are shown in Table 3.16. Since these are the
only NMR spectra available for any of the palladium(II)
complexes of N-substituted porphyrins, direct comparisons
with NMR parameters of corresponding non-bridged complexes
is impossible. However, comparisons can be made with the

free base forms of the N,N'-bridged porphyrin ligands (Table 3.12) and the mercury(II) complexes. The characteristic symmetry (two sets of paired doublets) for the pyrrolenine protons is evident for these bridged complexes. In the case of the palladium(II) complex where the bridging moiety has the benzyloxy side chain and two bromine atoms bound to the palladium atom, the mirror plane symmetry exhibited in the NMR spectrum is demonstrated in the solid state structure as well (the reflection plane bisects the Br–Pd–Br angle and includes the N–C, H and O atoms of the bridge). A comparison of the NMR shift for the proton bound to the bridging carbon atom of the free base (Table 3.14) and the corresponding shift for each of the palladium(II) complexes shows that there is a shift of about –1 ppm with the methyl and ethyl side chains and –2.5 with the benzyloxy side chain. Callot and coworkers noted that their bridge was much more stable in the presence of palladium(II) than the others (the others eventually give PdTPP even in the solid state). In all cases, binding of dihalopalladium(II) to the bridged ligands leads to upfield displacements of one pair from each of the two sets of doublets (the least downfield doublet is shifted about –0.16 to –0.20 ppm and the second most downfield doublet is shifted by –0.13 to –0.16 ppm). Since each set of doublets arises from neighboring symmetry-related pyrrolenine rings (one set from the two rings bound by the bridge), it is somewhat surprising, both that the effect of the bridging palladium(II) is as great on the rings bearing the bridge as it is on the two rings to which it binds covalently, and that the shifts are all so small. It is quite possible that the structure of the free base is very similar to that determined by Callot and coworkers for the palladium(II) complex.

Table 3.16. ^{1}H NMR Shifts of N,N'-Bridged Porphyrin Complexes of Pd(II), Hg(II) and Ni(II)[a]

Compound	N-C$\underline{H}$	N-COC$\underline{H}$	N-COCC$\underline{H}$	pyrrolenine[b] phenyl
Cl$_2$Pd–N,N'–N'',N'''–CHOCH$_3$ TPP	–3.00	–1.76		8.18,8.73(4.8);9.09,9.18(6.0) 8.3–8.8
Cl$_2$Pd–N,N'–N'',N'''–CHOC$_2$H$_5$TPP	–2.95	–2.02	–2.35	8.16, 8.70(5.0);9.06,9.16(6.2) 7.9, 8.2–8.8
Cl$_2$Pd–N,N',–N'',N'''–CHOCH$_2$C$_6$H$_5$TPP	–3.40	–0.03	3.28(o), 5.53(m), 6.06(p) 8.13,8.65(4.9);9.08,9.16(5.4) 7.9, 8.3–8.8	
Br$_2$Pd–N,N'–N'',N'''–CHOCH$_2$C$_6$H$_5$TPP	–3.50	–0.02	3.30(o), 5.56(m), 6.08(p) 8.14,8.66(4.9),9.08,9.16(5.4) 7.9, 8.3–8.8	
Cl$_2$Hg–N,N'–N'',N'''–CHOC$_2$H$_5$TPP	–3.70	–2.0	–2.0	8.44,8.85(4.7);9.14,9.33(5.2) 7.9, 8.2–8.5

	N–H			
Ni–N–tosyl–aminoTPP	–0.05	tolyl; 2.3(–CH$_3$), 4.86(d,2H), 6.36(d,2H) 7.8(s,2H),8.86(br,4H),8.7(s,2H) 7.6–8.5		
Ni–N–p–nitrobenzoylTPP	–0.6	p-nitrophenyl: 3.44(d,2H), 6.36(d,2H) 8.11(s),8.92(d),9.16(d),8.89(s) 7.7–8.35		

a Relative to TMS. Taken in CDCl$_3$. Reference for Pd(II) complexes: 6. Reference for Ni(II) complexes: 24.
b The pyrroleninic proton resonances for the Pd(II) complexes appear as pairs of doublets. The coupling constants are given in parentheses. Other designations: s = singlet, d = doublet and br = broad.

The displacements of all resonance of the pyrrolenine protons and for the proton bound to the bridging carbon atom are considerably greater in the case of mercury(II) than palladium(II). It is not clear whether this effect is due to the direct effect of the greater radius of the mercury(II) atom (due to a greater volume of electron density) or an indirect effect such as the distortion of the porphyrin ring to optimize bonding of the larger atom. Our earlier discussion of the structural effects of metal atom size for mono-N-methylporphyrins (Table 2.2) supports rather regular changes of bond lengths with atomic size, but changes in the angles of the pyrrolenine rings with respect to the porphyrin plane are not highly regular.

The nickel(II) complexes entered in Table 3.16 represent a quite different type of bridge – one that has a reflection plane oriented like that in a mono-N-methylTPP complex (see Figure 3.8). Thus a pattern of 2 singlets and 2 doublets is expected in the pyrrolenine proton resonance region. It should be noted that these two complexes, in which a nitrogen atom forms a bridge, are diamagnetic whereas a bridged complex of nickel(II) with a carbon atom (of an ethoxy-carbonylmethylene moiety) forming the bridge is paramagnetic,[59] as are the nickel(II) complexes of N-CH$_3$TPP and N-CH$_3$OEP.[60] Relative to zinc(II), the nickel(II) complex shows upfield shifts of the aromatic tolyl resonances (by -0.77 ppm (ortho?) and -0.37 ppm (meta?)) and a 0.4 ppm downfield shift of a pyrroleninic singlet, probably that of the proton bound to the N-substituted ring. The shifts of the p-nitrobenzoyl bridged complex of nickel(II) in the pyrrolenine region are slightly but significantly downfield (0.3 ppm for the substituted ring protons and 0.1-0.2 ppm for the others).

Proton Magnetic Resonance Spectra of Paramagnetic Complexes. Balch, LaMar and Latos-Grazynski and their coworkers have assigned proton NMR resonances for paramagnetic iron(II), nickel(II) and cobalt(II) complexes of N-methylporphyrins.[55], [60], [61] In these cases, extensive use was made of deuterated and methylated derivatives as well as comparisons of line widths and intensities to determine the assignments.

The iron(II) complexes are of special interest because species of this type are likely to be produced as intermediates when N-substituted protoporphyrin IX is formed from cytochrome P-450 or hemoglobin (Chapters 6-8). Previous work by LaMar and his coworkers has clearly demonstrated that the hyperfine shift patterns of iron porphyrins are highly sensitive to oxidation state, spin, and axial ligation. The data in Table 3.17 show the sensistivity of the shifts to such parameters.[55] Variation of the axial ligand from chloride to bromide to iodide causes large changes in shifts, notably for the N-methyl and pyrroleninic protons. Variation in the line widths in the order Cl> Br> I was noted by the authors to be characteristic of high spin and intermediate spin iron complexes and Fe(N-CH$_3$TPP)Cl previously has been demonstrated to be high spin.[62] Although iron complexes with a high degree of dipolar contribution to the chemical shift show displacements in the same direction and of comparable magnitude with o-H> m-H> p-H, in this case (comparisons of the shifts for Fe(N-CH$_3$TPP)Cl in Table 3.17), so the authors concluded that there was not a high degree of axial anisotropy. This finding is consistent with previous results for other high spin iron(II) complexes.

Table 3.17 ^{1}H NMR Shifts of Iron(II) Complexes of
N-Methylporphyrins[a]

Compound	Chemical Shifts, ppm (Line Width, Hz)				
		phenyl			
	N-CH₃	ortho	meta	para	pyrrolenine
Fe(N-CH₃TPP)Cl	105	14.6(60)	9.4	8.3	41.9(97)
	(973)	6.5(89)	9.3	6.6	31.7(62)
		5.1(82)	8.5		−0.25(137)
		4.3(51)	6.8		−0.44(79)
Fe(N-CH₃TPP)Br	92	15.2			46.4(81)
	(733)	8.3			32.3(32)
		7.4			3.0(89)
		5.9			0.1(52)
Fe(N-CH₃TPP)I	68	16.4			54.7(56)
	(484)	9.4			33.9(38)
		8.8			
Fe(N-CH₃TTP)Cl	104	14.6	9.2	4.4	42.0
		6.2	9.1	3.0	31.1
		5.0	8.2		0.26
		4.4	6.6		0.12
Fe((N-CH=C(p-C₆H₄Cl)₂Cl	161	20.1(72)	10.7	8.4	48.8(124)
	(125)	3.1(93)	9.5	5.6	27.4(43)
		0.0(67)	7.5		7.9(145)
		−0.5(44)	5.3		

	N-CH₃	*meso*	methylene		methyl
Fe(N-CH₃OEP)Cl	121	6.7	30.2	16.2	7.3
	(1290)	−8.7	26.0	15.4	5.5
		(164)	22.5	10.5	2.05
			20.5	5.4	0.21
Fe(N-CH₃OEP)Br	103	10.5	33.5	19.2	10.0
	(922)	−3.5	26.6	16.7	7.3
		(128)	25.8	11.6	1.8
			23.3	5.9	0.31

[a] Taken in CDCl₃, Ref. 55.

A notable finding of the study of Balch and coworkers was that the relatively sharp N–CD$_3$ resonance for Fe(N–CD$_3$TPP)Cl occurs at –138 ppm at –50^0 C.[55] This value corresponds closely with that of the Fe(N–CH$_3$TPP)Cl, which occurs at –143 ppm at –50^0 C (and at –105 ppm at 23^0 C). The vinylic proton of the complex derived from DDT, Fe((N–CH=C(p–C$_6$H$_4$Cl)$_2$TPP)Cl, occurs at even higher field (–161 ppm at 23^0 C). The occurrence of such shifts far from the normal region of resonances from biological samples and the relative sharpness of the deuterium peaks (estimated to be about 40–120 Hz in a protein sample), led the authors to suggest that this peak could be used to detect iron(II) complexes of natural N–substituted porphyrins. They calculated that a millimolar concentration of protein containing such metallo-porphyrins should be detectable.

By means of chemical oxidation of iron(II) complexes of N–CH$_3$TPP and N–CH$_3$TTP, Balch and coworkers have obtained the proton NMR spectra of iron(III) complexes.[35] For these spectra, the methods of assignment used in reference 55 were also employed. It is quite evident that the pyrrolenine resonances for the iron(III) complexes are much broader than those of the corresponding iron(II) complexes. The number of peaks shows that the complexes have the same low symmetry (C$_s$) and the positions and line widths are consistent with a high spin state (S = 5/2). Likewise, the decrease in line-width for bromide as axial ligand rather than chloride is normal for high spin iron(III) porphyrins. The extreme downfield shifts of the N–methyl protons, as in the case of the iron(III) complexes,[55] indicates a σ–spin transfer mechanism and is a distinctive feature of these spectra. Deuteration of the pyrrolenine rings shows that the resonances of one of the four pyrroles is shifted about 100 ppm upfield (to 2.4 ppm for ^{2}H) relative to the others. This one is likely the N–substituted ring. Unfortunately, this resonance comes in a crowded region of the spectrum and the other distinct feature, the N–CH$_3$ resonance, is very broad (even the ^{2}H resonance of N–CD$_3$ is 900 Hz wide). The patterns of shifts and line widths for the iron(III) complexes in Table 3.18 as well as and for those of the N-CH$_3$OEP complex[35] are consistent with the structure determined for Fe(N-CH$_3$TPP)Cl.[62] It might be very interesting to extend this type of study to complexes of N-methyldeuteroporphyrin IX, which has a β-pyrroleninic hydrogen atom like those of the tetraphenylporphyrins. A direct comparison of nmr parameters of this proton could be made for the two types of complex. Since the structure of N-methyltetraphenylporphyrin complexes have been established by x-ray crystallography, the structure of a close analog to the naturally-derived N-alkylporphyrins could then be firmly established.

Table 3.18. ^{1}H NMR Resonances of Iron(III) Complexes of N-Substituted Porphyrins[a]

| Compound | Chemical Shifts, ppm (Line Width, Hz) | | | |
| | | phenyl | | |
	N-CH$_3$	meta	para	pyrrolenine
Fe(N-CH$_3$TPP)Cl$^+$	272[b]	17.4(160)	6.2(70)	128(1200)[c]
		16.2(280)	-0.4(100)	92(1000)
		14.8(150)		79(1000)
		12.8(150)		
Fe(N-CH$_3$TPP)Br+		20.1(60)	6.5(28)	131(500)
		18.3(60)	-1.4(30)	96(460)
		17.2(60)		76(480)
		13.9(60)		
Fe(N-CH$_3$TTP)Cl$^+$		18.8(200)	17.6(60)	128(1500)
		16.5(170)	8.1(5)	90(1000)
		15.5(160)		76(1200)
		13.4(150)		
Fe(TPP)Cl		16.5	5	115
		14.6		

a In CDCl$_3$ at -50^0 C, Ref. 35, b This is the ^{2}H resonance for Fe(N-CD$_3$TPP)Cl$^+$, c For ^{2} pyrrolenine, shifts (and line widths) are: 126(160), 91(150), 76(140) and 2.4(140).

Balch, LaMar and coworkers have also reported spectra for iron(III) complexes of porphyrins with a vinyl carbene ($C=C(C_6H_4Cl_2)$) inserted between the iron atom and a porphyrin nitrogen atom.[63] These paramagnetic complexes (S = 3/2) show resonances that are greatly displaced from those of diamagnetic species, but for these species there is no proton on the N-substituent that is close enough to the coordination site to produce shifts greater than 50 ppm upfield. The pyroleninic protons of one ring (N-substituted) show downfield displacements while the other three rings have resonances displaced upfield. Substitution of the axial ligand from chloride to bromide to iodide causes downfield shifts for all of the pyroleninic resonances. The same effect had been found for the iron(II) complexes of N-methylporphyrins.[55] The pattern of line width variations for the different halide ligands is that normally found for high spin iron(III) complexes. The C_s symmetry of these complexes is evident from the shifts of the tetraphenylporphyrin complexes. The authors also obtained data for the carbene-inserted iron(III) complexes of octaethylporphyrin and protoporphyrin IX dimethyl ester. The methylene resonances are clearly evident in the region of −14 to −35 ppm for the spectrum of the octaethylporphyrin complex and the vinyl resonances for the protoporphyrin IX dimethyl ester complex occur upfield from TMS (about 0 to −8 ppm). Since a synthetic sample of the protoporphyrin IX complex was used without HPLC separation, all four isomers were detected in the sample.

The NMR spectra of nickel(II) N-methylporphyrins have been interpreted by Latos-Grazynski.[60] By specific ortho and para substitution of the phenyl rings of Ni(N-CH$_3$TPP)Cl, he assigned the phenyl resonances. By comparing r^{-6} values (calculated from the crystal structure of Co(N-CH$_3$TPP)Cl,[60] which can be assumed to have a very similar structure to that of the nickel(II) complex) with line widths he assigned the other resonances. Interestingly, the substituted pyrrolenine resonances occur upfield (see Table 3.19) while the other three rings have the downfield displacements normally found for nickel(II) complexes. For the N-CH$_3$OEP complex, areas and line widths were used to make the assignments shown in Table 3.20. Latos-Grazynski has performed a similar analysis of the spectra of cobalt(II) complexes of N-methyltetraphenylporphyrin and N-methyloctaethylporphyrin.[61]

Table 3.19. ^{1}H NMR Resonances of Chloronickel(II) Complexes of N-Methyl-5,10,15,20-tetraarylporphyrins[a]

Assignment	(ΔH/H)[b]	(ΔH/H)$_{iso}$	(ΔH/H)$_{dip}$	(ΔH/H)$_{con}$
N-CH$_3$	177.8	182.0	-31.8	213.8
pyrrole H(2,3)	9.84	-18.12	7.72	-25.84
pyrrole H(7,8)	33.67	24.72	6.74	17.98
pyrrole H(8,17)	72.64	63.69	6.74	56.95
pyrrole H(12,13)	68.19	59.32	7.24	52.08
o-phenyl(1)[c]	8.44	-0.16	2.30	-2.46
o-phenyl(2)	11.40	2.80	4.76	-1.96
o-phenyl(1')	5.91	-2.57	1.70	-4.27
o-phenyl(2')	12.74	4.14	5.80	-1.66
m-phenyl(1)	10.39	2.39	1.41	0.98
m-phenyl(1')	10.84	2.85	1.96	0.89
m-phenyl(2')	11.23	3.23	2.21	1.02
m-phenyl(2)	9.14	1.14	1.34	-0.20
p-phenyl(1,2)	8.01	0.13	1.52	-1.67
p-phenyl(1,2)	6.92	-1.85	1.60	-3.34
p-CH$_3$(phenyl 1,2)	6.00	3.30	1.12	2.18
p-CH$_3$(phenyl 1,2)	4.69	1.99	1.12	0.87
m-CH$_3$(phenyl 1)	2.39	-0.31	0.87	-1.18
m-CH$_3$(phenyl 1')	3.61	0.91	1.34	-0.43
m-CH$_3$(phenyl 2)	2.78	0.08	0.81	-0.73
m-CH$_3$(phenyl 2')	4.35	1.65	1.56	0.09
p-OCH$_3$(phenyl 1,2)	5.19	1.09	1.09	0.0
p-OCH$_3$(phenyl 1,2)	5.02	0.93	0.93	0.0

a Taken in CDCl$_3$, Ref. 60. Primed phenyl protons are on the same side of the porphyrin plane as the nickel(II) atom. Resonances for p-substituted *meso*-tetraaryl porphyrins are separated from each other.
b From Zn(N-CH$_3$T(p-RP)P)Cl and N-CH$_3$TH$_2$(p-RP)P[+] references. See Tables 3.6 and 3.13.

The temperature dependence of the shifts was used to estimate the dipolar contribution to the isotropic shift (see Table 3.19). In the case of the nickel(II) complexes, the contact and dipolar contributions are of the same magnitude. Since dipolar shifts for high spin nickel(II) complexes are usually small, Latos-Grazynski suggested that zero field splitting could be substantial and he showed that the temperature dependence seemed more consistent with a combination of zero field splitting and contact shift contributions.[60] However, the temperature dependence of all of the chemical shifts could not be fit with any simple model. The downfield shifts of the non-N-substituted rings in these complexes is consistent with σ-delocalization from the unpaired electron in the $d_{x^2-y^2}$ orbital of high spin nickel(II). He also suggested that the large upfield shifts at the *meso* positions indicate π electron density from metal-to-ligand backbonding through the overlap of the d_{yz} and d_{xz} orbitals. Arguing against a large degree of π backbonding, however, are the insensitivity of the visible absorption band intensities and positions to the number of d electrons and the occurrence of metal-halide stretching bands in infrared spectra at normal positions for M(II) rather than M(III) centers. Perhaps the spin density and net electron density at the metal atom are not closely related in these complexes.

Table 3.20. ^{1}H NMR Resonances for Chloro-N-methyl-octaethylporphinatonickel(II)[a]

	$(\Delta H/H)$	$(\Delta H/H)_{iso}$	$(\Delta H/H)_{con}$	$(\Delta H/H)$	$(\Delta H/H)_{iso}$	$(\Delta H/H)_{con}$
N-CH$_3$	190.2	194.8	228.6			
α-CH$_2$	38.95	34.55	25.95	23.57	19.57	14.97
	18.75	14.75	10.15	16.43	12.43	7.83
	13.90	9.90	5.30	11.73	7.73	3.13
	2.95	−1.05	−5.65			
β-CH$_3$	5.37	3.52	0.12	5.02	3.17	−0.23
	4.53	2.68	−0.72	3.87	2.04	−1.36
meso (5,20)				−15.18	−25.40	−35.48
meso (10,15)				−20.06	−30.36	−41.82

a Taken in CDCl$_3$ at −60° C. Ref. 60.

Carbon-13 NMR Spectroscopy. Very few ^{13}C data are available for N-substituted porphyrins. Johnson and Ward reported shifts for N-ethoxycarbonylmethyltetraphenylporphine, making the assignments; N-methylene at 43.4 ppm, ester methylene 59.7 ppm, ester methyl 13.2 ppm. The carbonyl carbon was reported at 166.6 ppm by its displacement relative to the multitude of peaks in the normal aromatic region (sixteen peaks were reported in the region from 119.8 to 157.8 ppm.[65] Assuming relatively rapid rotation of the phenyl rings and the existence of a reflection plane, there are 18 possible resonances.)

Assigned ^{13}C NMR data for the free base and zinc(II) complex of the compound are given in Table 3.21. The relatively low intensity expected for non-hydrogen-bearing sp^2 carbon bound to the nitrogen is evident in the spectra of the free base (Figure 3.10) and the zinc(II) complex (Figure 3.11). As expected, the ^{13}C resonance for this sp^2 carbon atom is far from the resonance observed by Johnson and Ward for an sp^3 carbon. Coordination of the 2-oxo group leads to signficant displacement of its ^{13}C resonance (assuming that the more displaced resonance is that of the 2-oxo carbon, 1.3 ppm for the zinc(II) complex and 5.9 ppm in the case of the bromocobalt(III) complex). A similar magnitude of displacement is observed for the N-C carbon conjugated to the 2-oxo carbon (1.4 ppm for the zinc(II) complex and 8.7 ppm for the bromocobalt(III) complex). One of the α-pyrrolenine resonances is clearly much different from the others (>162 ppm vs. 146.6 to 154.0 ppm), as is one of the β-pyrrolenine resonances (in the opposite direction, perhaps due to the resonance delocalization of N-substituted porphyrins). For the free base and the zinc(II) complex, one of these resonances appears 15 ppm upfield from the others. For the the bromocobalt(III) complex, the resonances of the *meso*-carbons are significantly displaced. Changes in resonances for the *meso*-phenyl carbon atoms and the other carbon atoms of the ligand are relatively small.

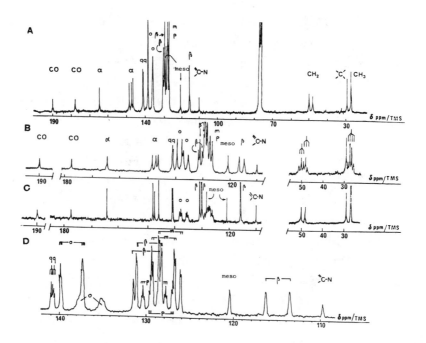

Figure 3.10. ^{13}C Spectra of $(N-C=C(OH)CH_2C(CH_3)_2CH_2CO)HTPP$,[76] A: decoupled, B: coupled, C: gated decoupled, D: expanded.

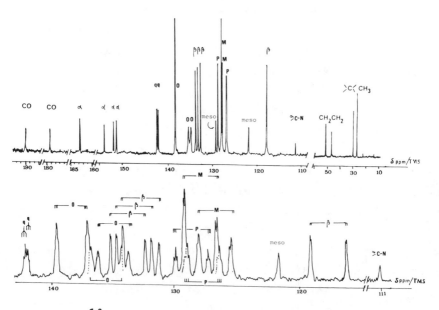

Figure 3.11. ^{13}C Spectra of $Zn[(N-\underset{76}{C}=C(O)CH_2(CH_3)_2CH_2CO)TPP]$. Upper: decoupled, lower: expanded.

Table 3.21. ^{13}C NMR Characteristics of Some N-Substituted Porphyrins.[a]

Assignment	Free Bases		Complexes	
	H-20	d-20	Zn(II)	BrCo(III)
	Porphyrin Ring			
α-pyrrolenine	164[b]	164.3	166.5[b]	162.6
	148.6[b]	148.2	154.0[b]	152.4
	147.4[b]	147.6	152.0[b]	149.3
	146.6[b]	146.1	151.2[b]	148.8
β-pyyrolenine	130.0(175)	130.7	133.8(175)	134.7
			133.3(175)	134.4
	129.7(175)	130.0	132.7(175)	128.3
	115.0(175)	115.6	117.7(176)	
meso	129.4	128.6	129.2	125.5
	120.5	120.7	121.7[c]	124.8
phenyl(1)	141.0[b]	141.1	142.2[d]	141.0
	140.8[b]	140.8	141.9[d]	140.4
phenyl(2,6)	138.6(162)	137.9[c]	138.8(161)	138.6
	136.6(162)	135.5[e]	135.5(161)	136.0
			134.9(161)	135.5
phenyl(3,5)	128.0(161)	[f]	127.9(161)[g]	128.3
	127.4(160)		127.6(160)[g]	127.9
phenyl(4)	129.1(160)[d]	[f]	128.6(160)[d]	127.9
	128.3(160)[d]		127.7(160)[d]	127.6
	N-Substituent			
C-O	190.2[h]	188.4	190.2[i]	188.7
	177.9[h]	177.4	179.2[i]	183.8
CH$_2$	49.9[i]	50.1	50.9[i]	50.5
	47.8[i]	48.1	46.2[i]	44.2
q-C	28.9	29.1	29.8	30.2
CH$_3$	26.8[k]	26.9	26.8[c]	26.3
N-C	109.7	109.9	111.1	118.4

a   The N-substituent is 2,4-dioxo-4,4'-dimethyl-cyclohex-1-ene. In the free base, the 2-oxo group is protonated. In complexes, it is bound to the metal atom. Spectra taken in CDCl$_3$, 0.01 M solutions, 20^0 C. Shifts in ppm from TYMS, ^{1}J in Hz in parentheses, Ref. 75.
b   Triplet, ^{2}J=8 Hz. For the others, ^{2}J or ^{3}J in Hz, d= doublet, t=triplet, q=quadruplet: c. t,3; d. 7,7; e. t, 22; f.126-129 ppm; g. d,7; h. t,2;i. t,120; j. t,128; k. q,126.

The [15]NMR Characteristics of N-Substituted Porphyrins.
Ogoshi has studied the ^{15}N NMR spectra of N-methyl derivatives of octaethylporphyrin.[66] At ambient temperature, the
exchange of a proton from one pyrroleninic nitrogen to
another is not as rapid in the case of N-methyloctaethylporphyrin as it is for octaethylporphyrin itself. As shown in
Table 3.22, the resonance for the proton-bearing and non-
proton-bearing pyrroleninic nitrogen atoms of octaethylporphyrin is averaged at 28° C (they are separate at -53° C)
but there are separate signals for N-methyloctaethylporphyrin. We have previously reported that the proton bound to a
free base N-methylporphyrin (N-methyltetrakis(p-bromophenyl)-porphine) is located on the ring opposite the methylated one in the solid state and is not found as partial
occupancy on the three non-alkylated nitrogens.[67] Ogoshi's
results suggest that it may be located similarly in solution
at ambient temperature, since two types of nitrogen are well
defined, each giving rise to one reasonably sharp resonance.
If the proton were located on ring adjacent to the methylated ring, there would be two types of unprotonated
nitrogens and two different uncoupled resonances. The
slower rate of internal hydrogen transfer is interesting in
view of Freeman and Hilbert's finding[68] that external proton
transfer in free base-cation equilibria of N-methylporphyrins is slower than for non-N-methylated porphyrins. The
spectrum of the monocation of the N,N'-dimethyl derivative
of octaethylporphyrin shows averaging at 28° C. Grigg and
coworkers have determined the structure in the solid state
of a related species, the monocation of N,N'-dimethyletio-
porphyrin I.[69] In that structure, the methyl groups are on
opposite sides of nitrogen atom formed. The two neighboring
unmethylated nitrogen atoms are displaced symmetrically 0.09
$\overset{\circ}{A}$ above and below the plane. The non-N-methylated rings
are nearly coplanar (each canted 4° from the nitrogen atom
plane). From the difference in the N-C-N angle (2.9°), the
authors suggested that the proton was localized in the solid
state. The thermal parameters of the two unmethylated nitrogen atoms, one protonated and the other not, are similar

to those of other atoms in the structure, suggesting an absence of disorder involving different hydrogen atom positions. Thus, the structure is highly symmetric. The energy of the two configurations of protonation would be the same and the distortion needed to allow proton transfer from one nitrogen to another should be small. Slow exchange of the free base of a mono-N-substituted porphyrin and rapid exchange for an N,N'-dimethylporphyrin monocation are in accord with the structural studies in the solid state.

In Ogoshi's data, the methylated nitrogen atom is clearly distinquished from protonated nitrogen atoms by a greater downfield displacement as well as by the absence of N-H coupling. The unprotonated nitrogen atom is, as expected, much more shielded than the others. In cases where other spectroscopic techniques are not definitive, ^{15}N NMR spectroscopy can provide a means of identifying the nature of N-substitution.

Table 3.22 ^{15}N NMR Characteristics of N-Substituted Octaethylporphyrins[a]

Compound	Solvent	Shifts, ppm (^{NH}J, Hz)		
		N	(N,NH ave.) NH	N-CH$_3$
H$_2$OEP	CDCl$_3$	139.7[c]	191(24)[b] 243.3(97)[c]	
H$_4$OEP^{2+}	TFA		253.7(93)	
N-CH$_3$HOEP	CDCl$_3$	125.8	243.3(100)	256.0
N-CH$_3$H$_3$OEP^{2+}	TFA		251.3, 2N(92) 252.9, 1N	259.0
N,N'-(CH$_3$)$_2$HOEP$^+$	CDCl$_3$		183.5(48)	256.3
N,N'-(CH$_3$)$_2$H$_2$OEP^{2+}	TFA		254.5	260.8

a Chemical shifts relative to NH$_4$NO$_3$ at 28° C unless otherwise specified. Ref. 66, b Exchanging rapidly, c At -53° C.

Electron Paramagnetic Resonance Spectroscopy

Few EPR data have been reported for N-substituted metal-loporphyrins and those few have been treated only qualitatively for the most part. The first EPR data for mono-N-alkylporphyrins were published in 1985 by Balch and coworkers,[35] Dolphin, Traylor and coworkers;[70] Latos-Grazynski and Jezierski;[71] and by my group.[72]

Data for copper(II) complexes of N-methyl- and N-p-nitro-benzyltetraphenylporphyrin were fit to values of $g_\parallel$ = 2.213, $g_\perp$ = 2.055 and $A_\parallel^{Cu}$ = 149 Gauss and $g_\parallel$ = 2.220, $g_\perp$ = 2.065 and $A_\parallel^{Cu}$ = 142 Gauss, respectively.[70] The EPR spectra of these complexes are quite different from that of CuTPP. The lower symmetry leads to a lack of resolution of superhyperfine structure. In addition, the $A_\parallel^{Cu}$ is smaller and the $g_\parallel$ value larger for the N-substituted complexes. The spectra are much better resolved in the presence of either the corresponding free base or the zinc(II) complex, indicating that these N-alkylated porphyrins aggregate in $CHCl_3$. Latos-Grazynski and Jezierski reported results for both copper(II) and manganese(II) complexes.[71] The parameters which they derived from their spectra (Table 3.23) are somewhat different from ours, but the same trends are clearly evident. In both cases, the authors interpreted the decrease in the $A_\parallel^{Cu}$ parameter to the geometry change of the copper(II) atom rather than to the effect of the weakened ligand field (which would be expected to increase $A_\parallel^{Cu}$). With regard to the parameters for $Mn(N-CH_3TPP)X$, the $g_\perp^{eff}$ region for all the solvents studied (Table 3.23) was 3.2 – 5.2 ppm and a weaker signal was also found in the $g_\parallel^{eff}$ = 2 region. In some , but not all cases, the hyperfine splitting from ^{55}Mn was evident. The results are consistent with a substantially non-axial geometry in all cases. As can be seen in Table 3.23, the values of $g_\perp^{eff}$ are strongly dependent on the nature of the axial ligand, with strong, nitrogen-containing ligands that most closely resemble porphyrin ligation giving the lowest values. Using di-n-butylamine, the authors were able to demonstrate the formation of an intermediate with this molecule axially coordinated to the manganese(II) atom, supporting a mechanism we have postulated for demethylation reactions[73] (see Chapter 4 for a discussion of these reactions.)

Table 3.23. EPR Parameters for Copper(II) and Manganese(II) Complexes of N-Methyltraphenylporphine.[a]

	Cu(N-CH$_3$TPP)X (CuTPP)		
	g$_\parallel$	g$_\perp$	A$_\parallel$, 10^{-4} cm^{-1}
Chloroform	2.256	2.080	172
	(2.190)	(2.055)	(212)
Pyridine	2.247	2.072	173
	(2.201)	(2.060)	(210)
Methanol	2.240	2.060	174
DBA/CH$_3$OH	2.243	2.060	174
(1:4 v/v)	(2.200)	(2.058)	(211)
N-methyl-imidazole	2.252	2.084	158
	Mn(N-CH$_3$TPP)X		
Chloroform	4.9–5.2		
N-methylimidazole	3.24		85
di-n-butylamine (DBA)	3.25		84
DBA/CHCl$_3$ (1:1 v/v)	3.80		84
DBA/CH$_3$CN (1:1 v/v)	4.05		
CH$_3$CN	4.15		
1,2-dimethylimidazole	4.15		85
pyridine	4.38		
DMSO	5.05		
DMF	5.10		
Collidine	5.20		

a At 130 K, Ref. 70.

The EPR characteristics of iron complexes of N-alkylporphyrins have been reported by Balch and coworkers,[35] and
Dolphin, Traylor and their coworkers.[70] Balch and coworkers
found that the iron(II) complexes of $Fe(N-CH_3TPP)X$, $Fe(N-CH_3TTP)X$ and $Fe(N-CH_3OEP)X$, where X = Cl or Br, did not
exhibit EPR signals in frozen chloroform solutions at 77 K.
Upon oxidation of $Fe(N-CH_3TPP)Cl$ with chlorine, however, an
EPR spectrum with $g_{\parallel}$ = 2.16 and $g_{\perp}$ = 5.86, characteristic of
high spin (S = 5/2) iron(II) porphyrins, was produced.
Likewise, oxidation of $Fe(N-CH_3TPP)Br$ produced an EPR spec-
trum with $g_{\parallel}$ =   2.14 and $g_{\perp}$ = 5.66 (also at 77 K in frozen
chloroform). The NMR spectra of these products are fully in
accord with this assignment and spin state.[35]

Dolphin, Traylor and coworkers used the EPR parameters of
intermediates in the interconversions of an N-2-hydroxy-4,4-
dimethylpentane porphyrin and a series of iron complexes to
assign the oxidation state of the iron as ferric.[61] The
intermediate, presumably six coordinate, with the oxygen
from the N-substituent bound to the iron atom on one side of
the porphyrin plane and a modified iodosylbenzene bound on
the other, gives a rhombic EPR spectrum typical of iron(III)
(g = 2.05, 5.35 and 8.57 in CH_2Cl_2 at 77 K), while addition
of pyridine to displace the modified iodosylbenzene moiety
gives an EPR spectrum typical of a low spin iron(III)
complex.

Mansuy and coworkers have presented EPR parameters of a
product from the reaction of DDT with $Fe(TPP)Cl$ to form a σ-
carbene, followed by oxidation.[74] This product consists of
a vinylidene group ($C=C(p-C_6H_4Cl)_2$) inserted between the
iron atom and one of the four pyrroleninic nitrogen atoms.
The EPR spectra were recorded from 4 K (where it is well
resolved) to 300 K (where the signals are broadened near g =
2 and less well resolved near g = 4). The EPR results are
consistent with magnetic susceptibility measurements that
indicate a "pure" intermediate spin state of ferric ion (S
= 3/2). The g values determined at 4 K are 2.03, 3.55 and
4.64. The 2.03 value is in perfect agreement with magnetic
susceptibility measurements.

X-Ray Photoelectron Spectroscopy

At present there is only one report of x-ray photoelectron parameters for N-substituted porphyrins.[75] In this report, the main comparison we made was between the values of the binding energies of the N 1s electrons of the methylated and nonmethylated nitrogen atoms (Figure 3.11). These binding energy differences are related to the difference in metal-nitrogen bond lengths determined crystallographically (Figure 3.12). Such a correspondence is not expected for non-N-substituted metalloporphyrins where the alteration in metal-nitrogen bond lengths is not as regular a function of the size of the metal atom and for which variable coordination numbers is a common occurrence. Comparatiuve values of x-ray photoelectron parameters for some N-methyltetraphenylporphyrin and for chlorins is given in Table 3.24. The position of the N 1s binding energy for the methylated nitrogen atom in the N-CH$_3$TPP complexes is quite similar to that for the 3,4-dihydropyrrolenine ring in chlorin complexes. The x-ray photoelectron technique is of limited utility for metalloporphyrins because the line widths (FWHM) are about 1 eV and differences between chemically distinct nitrogens, such as methylated, protonated or unsubstituted atoms is only about 0.2 eV. In addition, carbon atoms in the porphyrin ring or in simple alkyl or aryl N-substituents cannot be readily distinquished from one another. Advantages of the method include the requirement for only a very small amount of material (10^{-6} g) and its applicability to paramagnetic as well as diamagnetic complexes. In some cases, it may also provide a method of determining which atom of an axial ligand is bound to a metal atom.

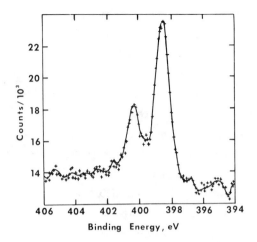

Figure 3.12. The N 1s region of the x-ray photoelectron spectrum of chloro–N–methyltetraphenylporphinatocobalt(II). The peak at higher energy is due to the methylated nitrogen and the other peak to the three unmethylated nitrogens.

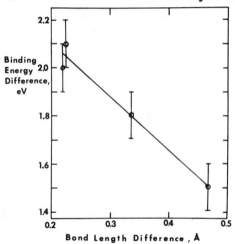

Figure 3.13. Correlation of the difference between the N 1s binding energies of the methylated and nonmethylated nitrogen atoms of the chloro–N–methyltetraphenylporphyrin complexes of Mn(II), Fe(II), Co(II) and Zn(II) (left to right) and the difference in metal–nitrogen bond lengths (the M–NCH$_3$ length minus the average of the other three M–N bonds.)

Table 3.24. X-Ray Photoelectron Spectroscopy Data (eV) For N-Methylporphyrins and Chlorins[a]

| | Metal Atom, M(N-CH$_3$TPP)Cl | | | | |
	Mn(II)	Fe(II)	Co(II)	Zn(II)	Free Base
Binding Energies					
sp^2 N 1s	398.3	398.2	398.4	398.5	397.6
sp^3 N 1s	400.4	400.2	400.2	400.0	399.9
Difference	2.1	2.0	1.8	1.5	2.3
Metal 2p$_{3/2}$	641.2	709.3	780.7	1022.2	
	Metal Atom, MTPC[b]				
	Co(II)	Ni(II)	Cu(II)	Zn(II)	Free Base
pyrrolenine 1s	398.0	398.3	398.1	397.6	397.4
dihydro-pyrrolenine 1s	399.7	399.0	399.7	398.5	399.3
Difference	1.7	0.7	1.6	0.9	1.9
Metal 2p$_{3/2}$	779.6	854.9	934.4	1021.3	

a   Ref. 75.
b   MTPC = *meso*-5,10,15,20-tetraphenylchlorin or *meso*-5,10,15,20-tetraphenyl-2,3-dihydroporphine.

Conclusion

Certainly the characteristic feature of a diverse variety of spectroscopic data for N-substituted porphyrins and metalloporphyrins is the effect of the lowered symmetry relative to corresponding non-N-substituted porphyrins. Although the visible absorption spectrum is generally a very good indicator of N-substitution, the ^{1}H NMR spectrum is generally easy to obtain for ligands or diamagnetic complexes and provides more definitive evidence. In addition, the symmetry evident in the pyrroleninic proton region readily distinguishes mono-N-substituted and N,N'-bridged species. For some paramagnetic complexes, the marked up-field shift of the protons (or deuterons) bound to the α-carbon atom of the nitrogen substituent provides a remarkably specific indication of N-substitution. It will be interesting to see results of this type for other metal ions. For those paramagnetic metal ions which obviate the use of NMR spectroscopy, similarities in visible absorption spectra with more well-defined diamagnetic complexes are perhaps the best indication of complexation in solution (along with evidence of fluorescence quenching by the paramagnetic ion). Removal of the metal ion and analysis of the remaining ligand, followed by reinsertion of the metal ion to obtain the same spectrum, is often employed to deduce the structure of paramagnetic complexes, but this method suffers from the ambiguity that has been demonstrated by the discovery of several bridged complexes (in which the N-substituent binds simultaneously to the metal atom and a nitrogen atom) which give the N-substituted free base when the metal is removed. Since there have not yet been any systematic studies of the infrared spectra of the bridged complexes, it is not clear whether this straightforward technique could be used to distinguish them from the more typical unbridged complexes.

In some cases, ^{15}N NMR and x-ray photoelectron spectroscopy may be useful to establish N-substitution (for diamagnetic species only in the former case), but their widespread use is unlikely. EPR is of limited utility except perhaps to deduce the oxidation state of iron complexes.

References

1. D.K. Lavallee and A.E. Gebala, Inorg. Chem., **19** (1974) 2004-2008.

2. D.K. Lavallee and M.J. Bain-Ackerman, Bioinorganic Chem., **9** (1978) 311-321.

3. D. Kuila, D.K. Lavallee, C.K. Schauer and O.P. Anderson, J. Am. Chem. Soc., **106** (1984) 448-450.

4. H.J. Callot and Th. Tschamber, Bull. Soc. Chim. Fr., **11** (1973) 3192-3198.

5. D.K. Lavallee, Coor. Chem. Rev., **61** (1985) 55-96.

6. H.J. Callot, J. Fisher and R. Weiss, J. Am. Chem. Soc., **104** (1982) 1272-1276.

7. H.J. Callot and E. Schaeffer, Tetrahedron Lett., **34** (1978) 2295-2300.

8. P.R. Ortiz de Montellano, H.S. Beilan and J.M. Mathews, J. of Medicinal Chem., **25** (1982) 1174-1179.

9. H.M.G. Al-Hazimi, A.H. Jackson, A.W. Johnson and M. Winter, J. Chem. Soc. Perkin I, (1977) 98-103.

10. M. Lange and D. Mansuy, Tetrahedron Lett., **22** (1981) 2561-2564.

11. J. Mercer-Smith, S. Figard, D.K. Lavallee and Z. Svitra, J. Nucl. Med., **26** (1985) 437 and in Antibody-Mediated Delivery Systems, Marcel-Dekker, New York, 1987.

12. F. De Matteis and L. Cantoni, Biochem. J., **183** (1979) 99-103.

13. H.J. Callot and Th. Tschamber, J. Amer. Chem. Soc., **97** (1975) 6175-6178.

14. F. De Matteis, and A.H. Gibbs, Biochem. J., **187** (1980) 285-288.

15. P.R. Ortiz de Montellano, P.R. Beilan and K.L. Kunze, Proc. Natl. Acad. Sci., U.S.A., **78** (1981) 1490-1494.

16. F. De Matteis, A.H. Jackson and S. Weerasinghe, FEBS Lett., **119** (1980) 109-112.

17. P.R. Ortiz de Montellano, G.S. Yost, B.A. Mico, S.E. Dinizo, M.A. Correia, and H. Kambara, Arch. Biochem. Biophys., **197** (1979) 524-33.

18. F. DeMatteis, A.H. Gibbs, P.B. Farmer, J.H. Lamb and C. Hollands, Adv. Pharmacol. Ther. Proc., 8th Int. Conf., Pergamon Press, New York, **5** (1982) 131-138.

19. F. DeMatteis, A.H. Gibbs and A.P. Unseld, in Biologically Reactive Intermediates 2: Chemical Mechanisms and Biological Effects, R. Snyder, et al., eds., Plenum Pub. Corp., N.Y., (1982) 307-340.

20. D.K. Lavallee, A. White, A. Diaz, J.-P. Battioni and D. Mansuy, Tetrahedron Lett., **27** (1986) 3521-3524.

21. M.J. Bain-Ackerman, Ph.D. Dissertation, Colorado State Univ., (1978).

22. W.K. McEwen, J. Amer. Chem. Soc., 68 (1946) 711-713.

23. A.H. Jackson and G.R. Dearden, Ann. New York Acad. Sci., 206 (1973) 151-176.

24. H.J. Callot, B. Chevrier and R. Weiss, J. Am. Chem. Soc., 100 (1978) 4733-4741.

25. T.J. Wisnieff, A. Gold and S.A. Evans, Jr., J. Am. Chem. Soc., 103 (1981) 5616-5618.

26. D.K. Lavallee, Bioinorg. Chem., 6 (1976) 219-227.

27. H. Ogoshi, S. Kitamura, H. Toi and Y. Aoyama, Chem. Soc. of Japan, Chem. Lett., (1982) 495-498.

28. H. Ogoshi, E.I. Watanabe, N. Koketzu and Z.I. Yoshida, J. Chem. Soc., Chem. Commun., (1974) 943-944.

29. R. Grigg, G. Shelton and A. Sweeney, J.Chem. Soc. Perkin Trans. I, (1972) 1789-1799.

30. D.K. Lavallee, J. Inorg. Biochem., 16 (1982) 135-143.

31. D. Kuila, A.B. Kopelove and D.K. Lavallee, Inorg. Chem., 24 (1985) 1443-1446.

32. D. Dolphin, D.J. Halko and E. Johnson, Inorg. Chem., 20 (1981) 4348-4351.

33. D. Kuila, Dissertation, CUNY, (1984).

34. D.K. Lavallee and D. Kuila, Inorg. Chem., 23 (1984) 3987-3992.

35. A.L. Balch, G.N. LaMar, L. Latos-Grazynski and M.W. Renner, Inorg. Chem., 24 (1985) 2432-2436.

36. K.M. Kadish, In Iron Porphyrins, Part 2, A.B.P. Lever and H.B.Gray, eds., Addison-Wesley, N.Y., 1983, 161-251.

37. H.J. Callot, Bull. Soc. Chim. Fr., 11 (1972) 4387-4391.

38. H. Ogoshi, T. Omura and Z.I. Yoshida, J. Chem. Soc., 95 (1973) 1666-1668.

39. D.K. Lavallee, T. McDonough and L. Cioffi, Applied Spectroscopy, 36 (1982) 430-435.

40. D.K. Lavallee and J. Andrew, <u>Analyt. Chem.</u>, **49** (1977) 1482-1485.

41. R. Becker, <u>Theory and Interpretation of Fluorescence and Phosphorescence</u>, Wiley, New York, (1969).

42. D.K. Lavallee, <u>Inorg. Chem.</u>, **17** (1978) 231-236.

43. G. Chottard, D. Mansuy and H.J. Callot, <u>Inorg. Chem.</u>, **22** (1983) 362-364.

44. L.J. Boucher and J.J. Katz, <u>J. Am. Chem. Soc.</u>, **89** (1967) 1340-1345.

45. L.J. Boucher, <u>J. Am. Chem. Soc.</u>, **90** (1968) 6640-6645.

46. D.W. Thomas and A.E. Martell, <u>J. Am. Chem. Soc.</u>, **81** (1959) 5111-5119.

47. J.W. Buchler, In <u>Porphyrins and Metalloporphyrins</u>, K.M. Smith, Ed., Elsevier, New York, NY (1975), 157-231.

48. H. Ogoshi, Y. Saito and K. Nakamoto, <u>J. Chem. Phys.</u>, **57** (1972) 4194-4202.

49. W.S. Caughey and P.K. Iber, <u>J. Org. Chem.</u>, **28** (1963) 269-270.

50. P.R. Ortiz de Montellano and K.L. Kunze, <u>J. Am. Chem. Soc.</u>, **103** (1981) 6534-6536.

51. C.K. Sauer, O.P. Anderson, D.K. Lavallee, J.-P. Battioni and D. Mansuy, <u>J. Amer. Chem. Soc.</u>, **109** (1987)

52. H.J. Callot and E. Schaeffer, <u>Tetrahedron Lett.</u>, **21** (1980) 1335-1338.

53. A.H. Jackson, <u>Porphyrins</u>, D. Dolphin ed., vol, Acad. Press, NY (1978) 341-364.

54. H.J. Callot and F. Metz, <u>J. Chem. Soc., Chem. Commun.</u>, (1982) 947-948.

55. A.L. Balch, Y-W. Chan, G.N. LaMar, L. Latos-Grazynski and M.W. Renner, <u>Inorg. Chem.</u>, **24** (1985) 1437-1443.

56. K.L. Kunze and P.R. Ortiz de Montellano, <u>J. Am. Chem. Soc.</u>, **103** (1981) 4225-4230.

57. K.L. Kunze, B.L.K. Mangold, C. Wheeler, H.S. Beilan and P.R. Ortiz de Montellano, <u>J. Biol. Chem.</u>, **258** (1983) 4202-4207.

58. P.R. Ortiz de Montellano, H.S. Beilan, K.L. Kunze and B.A. Mico, J. Biol. Chem., 256 (1981) 4395-4399.

59. H.J. Callot, Th. Tschamber, B. Chevrier and R. Weiss, Angew. Chem., Int. Ed., 87 (1975) 545.

60. L. Latos-Grazynski, Inorg. Chem., 24 (1985) 1681-1686.

61. L. Latos-Grazynski, Inorg. Chem., 24 (1985) 1104-1105.

62. O.P. Anderson, A.B. Kopelove and D.K. Lavallee, Inorg. Chem., 19 (1980) 2101-2107.

63. A.L. Balch R.-J. Cheng, G.N. LaMar and L. Latos-Grazynski, Inorg. Chem., 24 (1985) 2651-2656.

64. O.P. Anderson and D.K. Lavallee, J. Amer. Chem. Soc., 99 (1977) 1404-1409.

65. A.W. Johnson and D. Ward, J. Chem. Soc. Perkin I, (1977) 720-723.

66. K. Kawano, Y. Ozaki, Y. Kyogoku, H. Ogoshi, H. Sugimoto and Z.I. Yoshida, J. Chem. Soc., Perkin II, (1978) 1319-1325.

67. D.K. Lavallee and O.P. Anderson, J. Amer. Chem. Soc., 104 (1982) 4707-4708.

68. K.A. Freeman and F. Hibbert, J. Chem. Soc., Perkin II, (1979) 1574-1578.

69. A.M. Abeysekera, R. Grigg, J. Trocha-Grimshaw and K. Hendrick, Tetrahedron, 36 (1980) 1857-1868.

70. T. Mashiko, D. Dolphin, T. Nakano and T.G. Traylor, J. Amer. Chem. Soc., 107 (1985) 3735-3736.

71. L. Latos-Grazynski, Inorg. Chim. Acta., 106 (1985) 13-18

72. W.V. Sweeney, D. Kuila and D.K. Lavallee, Inorg. Chim. Acta., 99 (1985) L9-L11.

73. D.K. Lavallee, Inorg. Chem., 16 (1977) 955-957.

74. D. Mansuy, I. Morgenstern-Badarau, M. Lange and P. Gans, Inorg. Chem., 21 (1982) 1427-1430.

75. D.K. Lavallee, J. Brace and N. Winograd, Inorg. Chem., 18 (1979) 1776-1780.

76. J.-P. Battioni, I. Artaud, D. Dupré, P. LeDuc and D. Mansuy, unpublished results.

CHAPTER 4

CHEMISTRY OF N-SUBSTITUTED PORPHYRINS III:
REACTIONS OF N-SUBSTITUTED PORPHYRINS AND METALLOPORPHYRINS

Introduction

The few studies of metal complexes of N-substituted porphyrins reported before 1975 mostly concerned synthesis and
spectroscopic properties, with the exceptions of two important articles by Hambright and coworkers who reported
investigations of the kinetics of formation of zinc and
cadmium N-methyletioporphyrin III[1] and the hydrolysis of
zinc N-methyletioporphyrin III.[2] These studies indicated
that the rates of reactions of the N-substituted porphyrins
were very different from those of corresponding non-N-substituted porphyrins. Together with their observation that
the methyl group of N-methyletioporphinatozinc(II) is lost
on heating in pyridine,[3] showing that carbon-nitrogen bond
cleavage is more facile than normal for these compounds,
these articles have provided the impetus for several subsequent kinetic and mechanistic studies by our group and by
others. Another body of work by the groups of Ogoshi,
Dolphin and Callot has been concerned with the mechanisms of
reactions involving decomposition of N-alkylporphyrin complexes to form either σ-complexes or homoporphyrins. These
reactions are of interest with respect to biological processes and synthesis, respectively.

This chapter will treat the two most general reactions of
free base N-substituted porphyrins, protonation and metal
complexation, and those of the metal complexes, hydrolytic
demetalation and oxidation-reduction processes, as well as
several reactions which are specific to the N-substituted
metalloporphyrins: nucleophilic displacement of the N-substitutent, migration of the N-substitutent to the metal atom
and homoporphyrin formation.

Acid–Base Equilibria of N–Substituted Porphyrins

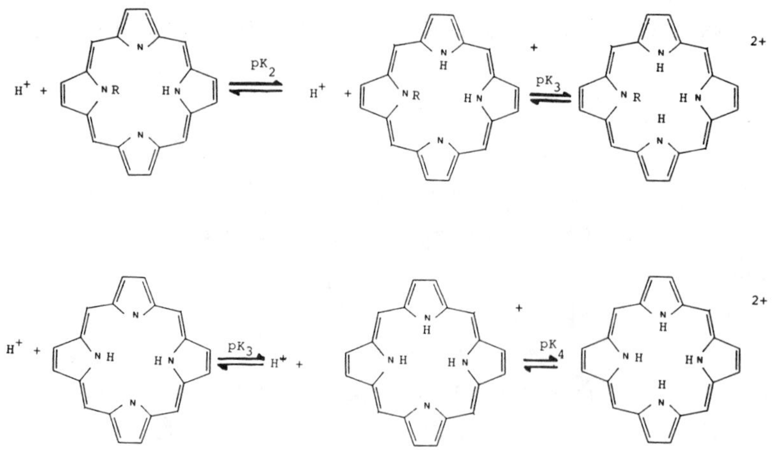

Figure 4.1. Acid–base equilbria of N–substituted (upper) and non–N–substituted (lower) porphyrins.

Mono–N–Substituted Porphyrins. The first publication mentioning an N–substituted porphyrin, that of McEwen in 1936[4] reported a pK_a value of 14–15 and compared its value to etioporphyrin I. At that time, etioporphyrin I was used as an acid–base indicator for very weak acids and the color change associated with the transformation from the sodium salt to the free base was used as a titration end point indicator. McEwen noted that the two pK's for etioporphyrin I are close to one another and, therefore, use of a dye with a single protonation equilibrium such as N–methyletioporphyrin I would make interpretation easier. He made the assumption that the color change for N–methyletioporphyrin I corresponded to the monoanion–free base transition. Subsequent studies of the acid–base equilibria of N–substituted porphyrins have involved, instead, the equilibria between free base, mono- and diprotonated forms. In 1952, Neuberger and Scott reported pK values for the free acid and esterified forms of N-methylcoproporphyrin I.[5] The difference in the pK value for the free base-monocation equilibrium (pK_3 for the non-N-substituted porphyrins and pK_2 for N-substituted porphyrins, see Figure 1) and the pK value for the monocation-dication equilibrium (pK_4 for non-N-substituted porphyrins and pK_3 for N-substituted porphyrins) were much greater in the case of the N-methylporphyrin than coproporphyrin itself (11.3 vs. 0.7 and 8.3 vs 0.7, compared with 7.2 and 4.2). They also commented, of course, on the greater basicity of the N-methylated porphyrins. Observations of naturally derived porphyrins, such as N-alkylated protoporphyrin IX, deuteroporphyrin IX, hematoporphyrin IX, etc., and non-*meso*-substituted porphyrins such as octaethylporphyrin,[6] and the etioporphyrins indicate that they are also stronger bases than the corresponding non-N-alkylated porphyrins; also, their pK_2 and pK_3 values are more widely separated than the pK_3 and pK_4 values of corresponding non-N-substituted porphyrins. However, these observations are qualitative and few quantitative data exist.

The basicity of the most widely used synthetic porphrin, H_2TPP, does not change as drastically when it is alkylated (pK_2 and pK_3 values are 5.6 and 3.9 for N-CH_3HTPP in nitrobenzene and the pK_3 and pK_4 values are 4.4 and 3.9 for H_2TPP[7]), but the difference in the values for the successive protonations is still great enough that the monocationic form can be clearly distinguished spectroscopically. Freeman and Hibbert have reported a careful comparison of N-CH_3HTPP and H_2TPP in a mixture of DMSO and water in which they obtained data both for the pK values and for the rates of interconversion of the mono- and diprotonated forms.[8] They found that the difference for pK_2 and pK_3 of N-CH_3HTPP in 90% (v/v) DMSO/H_2O was 3 whereas it is nearly zero for H_2TPP. The relaxation time for the equilibrium of N-$CH_3H_2TPP^+$ and N-$CH_3H_3TPP^{2+}$ was outside the range of their temperature jump method but the slower equilibrium for the corresponding p-methoxyporphyrin (which is more basic, with

$pK_4 = 2.6$ for N-CH$_3$HTp-OCH$_3$P vs. 1.4 for N-CH$_3$HTPP) provided a rate constant of 4.3×10^5 M^{-1}s^{-1} for protonation of the monocation, a relatively slow reaction. The corresponding reaction for non-N-substituted tetrakis(p-methoxyphenyl)-porphine is considerably faster (outside the range of the experiment). Earlier, Johnson and Jackson and their coworkers had shown by nmr spectroscopy that the interconversion of free base N-CH$_3$HTPP and the monoprotonated form, N-CH$_3$H$_2$TPP^{2+} is relatively slow,[9] so it appears that both steps in the ionization of the diprotonated forms of N-methyltetraarylporphyrins are relatively slow processes. The contrast with the rapid rate of deprotonation of non-N-substituted porphyrins may be due to the fact that their deprotonation leads to a more stable planar free base but, in the case of N-substituted porphyrins, the free base is still nonplanar. As Freeman and Hibbert noted, the difference in pK_2 and pK_3 values for the N-methyltetraarylporphyrins is about the magnitude expected for dibasic organic compounds and it is the non-N-substituted porphyrins which are unusual in this respect. From this point of view, however, it is not evident why there is such a great difference between the pK_2 and pK_3 values of the non-*meso*-substituted N-substituted porphyrins.

Multiply-N-Substituted Porphyrins. Although there are no quantitative data for the acid-base equilibria of N,N'- and N,N''-dialkylated porphyrins, both the references to the behavior of these compounds in the literature and our experience with them have shown no exceptions to the rule that they are highly basic and significantly more so than the corresponding mono-N-alkylporphyrin. Jackson and Johnson and coworkers compared the behavior of *trans*-N,N'-dimethyltetraphenylporphyrin and the *trans*- and *cis*-dimethyloctaethylporphyrins, showing that it was possible to isolate pure samples of the free base form of the former but not of the latter compounds. Thus, it appears that the dialkyl derivatives of the *meso*-tetraarylporphyrins are weaker bases than dialkyl derivatives of other porphyrins.

Complexation of N-Substituted Porphyrins

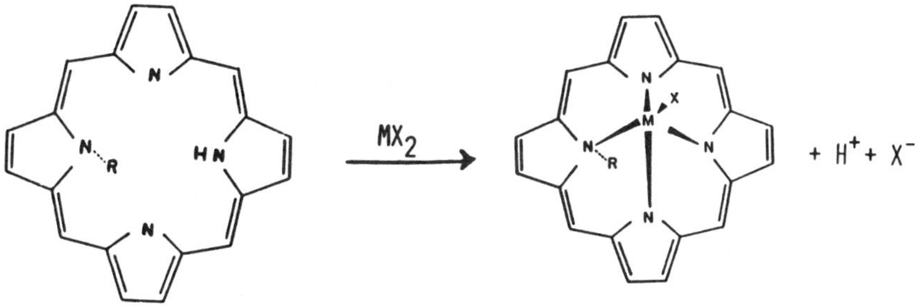

Figure 4.2.   Metalation of an N-substituted porphyrin.

Background.   In 1946, McEwen reported the first complex of an N-substituted porphyrin, N-methyletioporphyrin I zinc(II)[10]. (He also reported the copper(II) complex, but in view of the reported spectroscopic properties and the fact that he reported his complex to be stable toward concentrated hydrochloric acid, it is only reasonable to conclude that he had isolated the copper complex of demethylated etioporphyrin I). The first report of the kinetics of complexation (metalation) of an N-alkylporphyrin was by Hambright and coworkers, who showed that the formation of the zinc(II) and cadmium(II) complexes of N-methyl-etioporphyrin III in dimethylformamide were about 10^5 faster than the reactions with etioporphyrin III.[1] Since that study, our group and Tanaka's have systematically investigated the metalation kinetics of N-methyltetraphenyl-porphyrin using several different metal ions. These data are consistent with a dissociative interchange mechanism in which the rate of loss of a ligand in the first coordination sphere of the precursor metal complex determines the order of reaction rates.

Kinetic Trends. The first systematic study of metalation reactions of N-substituted porphyrins was published by Bain-Ackerman and Lavallee in 1979.[11] In this study, complexes were prepared in which the first coordination sphere about each metal ion consisted of six dimethylformamide molecules, with perchlorate as the counterion (for safety, such complexes should be prepared using an "innocent" counterion that is less dangerous, such as trifluoromethanesulfonate or tetrafluoroborate) so that there would be no question about the nature of the metal ion complex that actually reacts with the porphyrin when dimethylformamide is used as solvent. In prior studies of the metalation of porphyrins, use of chloride or acetate salts which undergo ligand exchange in polar solvents such as dimethylformamide to yield a variety of possible metal ion precursor complexes, had presented problems of interpretation.[12] Under pseudo-first-order conditions with the metal salt in excess, the reactions were all strictly first order with respect to the N-CH$_3$HTPP concentration and also showed a first order dependence on metal ion concentration. Such a well-behaved second order rate law has also been found for the reactions of chloride salts with H$_2$TPP in dimethylformamide[13] and for reactions of hexaaqua perchlorate salts with tetrakis(4-N-methylpyridyl)porphine in aqueous solution,[14] but it is not typical of metalation reactions of non-N-substituted porphyrins.[12] In fact, reactions of nitrate salts with H$_2$TPP in dimethylformamide show more complex behavior.[15]

Hambright and Chock had previously shown that the order of the rates of reaction of aquo salts of Cu(II), Zn(II), Co(II), Mn(II) and Ni(II) correlated well with the order of the rates of solvent exchange of these metal ions, with the sole exception being Mn(II), whose complexation rate was slower than expected on the basis of its solvent exchange rate.[14] We found that the same order holds for reactions of N-CH$_3$HTPP in dimethylformamide (Table 4.1).[11]

Table 4.1 Comparison of Rate Constants for N-Methyltetra-phenyl- and Tetrakis(4-N-methylpyridyl)porphine Metalation

N-methyltetraphenylporphine in DMF[a]

k, $M^{-1}s^{-1}$ (25°C)	normalized rate constant	k_d, DMF, s^{-1}	$10^6 k/k_d$, M^{-1}
Cu(II)	290	9.6×10^5	2.5×10^8	1.15
Zn(II)	10	3.5×10^4	3×10^6	3.3
Co(II)	0.68	2.3×10^3	2.5×10^5	2.7
Mn(II)	0.010	33	4.0×10^6	0.0027
Ni(II)	0.0003	1	4.0×10^3	0.075

Tetrakis(4-N-pyridyl)porphine in H_2O[b]

k, $M^{-1}s^{-1}$ (25°C)	normalized rate constant	k_d, DMF, s^{-1}	$10^6 k/k_d$, M^{-1}
Cu(II)	2.3	4.6×10^3	2.5×10^9	0.86
Zn(II)	0.050	1.0×10^3	3.2×10^7	1.58
Co(II)	0.0022	43	2.5×10^6	0.86
Mn(II)	0.0025	50	3.2×10^7	0.08
Ni(II)	0.00005	1	2.8×10^4	1.78

a Ref. 11
b Ref. 14

Metalation Reaction Mechanism. These results are consistent with a multistep mechanism (Scheme 4.1), first proposed by Hambright and Chock,[14] in which ligand dissociation
from the metal atom is an important step in determining the
overall rate of the reaction. The exceptional rate for
manganese(II) has been attributed to an equilibrium in which
the formation of the first bond between the metal atom and a
pyroleninic nitrogen atom is reversible.[12] Since high spin
manganese(II) has no ligand field stabilization energy, it
has the least affinity for a nitrogenous ligand relative to
an oxygen atom donor solvent and, hence, this reversible
step would be the least favorable of all the metal ions in
the series studied. This interpretation is consistent with
acid dissociation studies, which show that cleavage of the
last metal–nitrogen bond during demetalation is relatively
rapid (as mentioned above). Since there was no difference in
the overall rate (in the absence of water) for reactions
using N-CH$_3$DTPP[11] and D$_2$TPP[16] and the corresponding
protonated species, we concluded that the loss of protons
(the final step in Scheme 4.1) is not rate determining for
either the N–substituted or the non–N–substituted porphyrin
metalation reactions. Tanaka has reported the same
finding.[15]

Three differences in the overall mechanism (Scheme 4.1)
between N–substituted and non–N–substituted porphyrins are:
1) there is probably no deformation step required for the
"predeformed" N–substituted porphyrins (as discussed in
Chapter 2, the structure of the porphyrin ring of the free
ligand and of metal complexes of N–substituted porphyrins
are very similar), 2) the N–substituted porphyrins are more
polar, which should make outer–sphere complexation with
positively charged metal ions more favorable, and 3) only
one proton is lost in the final step (the N–substituted
porphyrins act as monovalent anionic ligands while the non–
N–substituted porphyrins are divalent).

N-Substituted Porphyrin   **Non-N-Substituted Porphyrin**
                                                      Deformation

''Predeformed''

$$H_2P \quad \rightleftharpoons \quad H_2P^*$$

Outer Sphere Complexation

$$ML_6{}^{2+} + N-RHP \rightleftharpoons [L_6M,N-RHP]^{2+}$$

$$ML_6{}^{2+} + H_2P \rightleftharpoons [L_6M,H_2P^*]^{2+}$$

Ligand Dissociation and First Bond Formation

$$[L_6M,N-RHP]^{2+} \rightleftharpoons L_5M-N-RHP^{2+} + L$$

$$[L_6M,H_2P^*]^{2+} \rightleftharpoons L_5M-H_2P^{2+} + L$$

Second Bond Formation

$$L_5M-N-RHP^{2+} \rightleftharpoons L_nM=N-RHP^{2+} + (5-n)L$$

$$L_5M-H_2P^{2+} \rightleftharpoons L_nM=H_2P^{2+} + (5-n)L$$

Final Step – Proton Dissociation

$$L_nM=N-RHP^{2+} \rightleftharpoons LMN-RHP^+ + L_{n-1} + H^+$$

$$L_nM=H_2P^{2+} \rightleftharpoons MP + L_n + 2H^+$$

Scheme 4.1

Volumes of Activation. The finding that the order of
reaction rates for a variety of metal atoms is related to
ligand dissociation rates is certainly consistent with a
rate determining step that is dissociative. But it is not
definitive. Tanaka has used a high-pressure stopped-flow
instrument to determine the activation volumes for metalation reactions of N-methyltetraphenylporphyrin in DMF.[17,18]
The rate constants he determined were very similar to our
values with the exception of the reaction of Mn(II) (this
value still, however, reflects its anomalous position in the
order of reaction rates as discussed above). In his case,
the precursor complexes were nitrate rather than perchlorate

salts and it may be that Mn(II) has a significant affinity for nitrate ion in DMF. The reaction volumes of the reactions are all positive (Table 4.2), conclusively demonstrating that the rate determining step is dissociative.

Table 4.2   Rate Constants and Volumes of Activation for Metalation of N-Methyltetraphenylporphyrin[a]

M(II)	k, 25°C $M^{-1} s^{-1}$	$\Delta H^{\ddagger}$ kcal/mol	$\Delta S^{\ddagger}$ eu/mol	$\Delta V^{\ddagger}$ cm^3/mol
Hg(II)	$(7 \pm 3) \times 10^5$	5.5 ± 0.7	-14 ± 2	4.3 ± 0.8
Cd(II)	580 ± 10	13 ± 0.7	-3 ± 2	8.9 ± 1.6
Zn(II)	10 ± 0.8	14 ± 0.7	-7 ± 3	7.0 ± 0.6
Co(II)	0.38 ± 0.3	17 ± 0.3	-3 ± 1	8.0 ± 0.3
Mn(II)	0.14 ± 0.01	19 ± 0.1	-0.6 ± 0.4	12.9 ± 0.8

a   Ref. 18.

Comparisons With Corresponding Non-N-Substituted Porphyrins. Since the N-substituted porphyrins are "predeformed", the first preequilibrium step is avoided and the reaction can proceed more rapidly. The first study of the acceleration of metalation by N-substitution involved the complexation of cadmium(II) and zinc(II) to etioporphyrin and N-methyletioporphyrin in dimethylformamide.[1] Hambright and his coworkers found that the reactions of the N-methyl-etioporphyrin are about 10^5 times faster. More recently, Tanaka and coworkers have studied the reactions of H_2TPP and N-CH$_3$HTPP with cadmium(II), copper(II), and zinc(II) under identical conditions in dimethylformamide. As in our case,[11] they found that the reactions with N-CH$_3$HTPP were strictly first order with respect to metal ion concentration. In addition, they found that while the reactions of H_2TPP with cadmium(II) were first order in metal ion concentration throughout the range studied, for copper(II) and zinc(II) they were first order only at low ($[M(II)] < 10^{-4}$ M) or high concentrations ($[M(II)] > 10^{-2}$ M) with a behavior in the intermediate region that was consistent with formation of an intermediate that could be converted to the metalloporphyrin

product with the assistance of a second metal ion. Presumably, this second metal ion would attack from the opposite
face of the porphyrin, an impossible mechanism for the N-
methylporphyrin (hence, the explanation of the difference in
behavior).   If the effective rate constants in the low
concentration region are compared, the ratios of rate constants for N-methyltetraphenylporphyrin and the tetraphenylporphyrin reactions are 250 for copper(II) and 10^3 for
zinc(II). At high concentrations the ratio is 10^4 for both
copper(II) and zinc(II). The ratio of rate constants for
cadmium(II) is about 10^5 throughout the range studied.

For non-N-substituted porphyrins, basicity (as reflected
in pK_3 values) may be an important factor in determining the
differences in rates of different porphyrins.   Correlations
of rate with basicity have been reported that indicate that
there is a nearly linear increase in metalation rate with
increasing pK_3 values.[12]   Nearly all of the kinetic data
for metalation of N-substituted porphyrins have involved N-
methyltetraphenylporphyrin, but there are a few data
available for other N-substituted porphyrins that bear on
the question of the affect of basicity on rate. The ratios
of rate constants for N-methyltetraphenylporphyrins and
tetraphenylporphyrin found by Tanaka at high concentration
are quite similar to the ratio for the more basic N-methyletioporphyrin III and its non-N-substituted analog,
etioporphyrin III, differing by about a factor of 10 for
zinc(II). The absolute magnitudes of the rate constants for
N-methyltetraphenylporphyrin and N-methyletioporphyrin III
at 25° C are 10 M^{-1} s^{-1} and 520 M^{-1} s^{-1}, respectively, for
Zn(II) and 580 M^{-1} s^{-1} and 1150 M^{-1} s^{-1} for cadmium(II).
The difference in basicities of the non-N-substituted
porphyrins, tetraphenylporphyrin and etioporphyrin III, is
significant but not as great as for the N-methylated
analogs. The rate constants at 25° C in the high concentration region for reactions of zinc(II) are 9 x 10^{-4} M^{-1} s^{-1}
and 4.7 x 10^{-3} M^{-1} s^{-1} and for cadmium(II) they are 6.5 x
10^{-3} M^{-1} s^{-1} and 17 x 10^{-3} M^{-1} s^{-1}.[11,15] Thus, the
differences in the rate constants for the N-methylporphyrins

are similar to those for the less basic non–N–methylated analogs. Hambright and his coworkers have derived a quantitative relationship between metalation rates and porphyrin basicity.[19] However, it does not appear that the relation they derived porphyrins can be applied directly to reactions of N–substituted porphyrins.

Since the pK_3 and pK_4 values of non–N–substituted porphyrins are highly correlated to each other (and perhaps even to the removal of a proton, that is, to their pK_2 values), the source of the relationship between metalation rate and basicity may be subtle. It is not surprising that a different relation may hold for the N–substituted porphyrins. Since the kinetic studies to date involve only two N–substituted porphyrins, a quantitative estimate of the effect of basicity on rate or its fundamental origin is premature: a wider range of experimental data must be obtained before any firm conclusions can be drawn.

Practical Consequences. Because a predominant feature of the reaction mechanism for metalation of N–substituted porphyrins is dissociation of ligand from precursor metal complex, metalations are best carried out using monodentate complexes in solvents that exchange readily, such as alcohols or organonitriles. From the limited kinetic data now available and observations during synthesis of complexes of a wide variety of N–substituted porphyrins, there does not seem to be a great difference in rate for *meso*– and non-*meso*–substituted porphyrins, so conditions appropriate for one type of porphyrin (keeping in mind their different basicities) are generally adequate for others. One limitation with respect to solvent is the dissociation of the N–substituent in the presence of strong nucleophiles. Loss of the alkyl group is much more facile after the metal complex has been formed (dealkylation is discussed in Chapter 5).

Tanaka and coworkers have suggested that the very rapid rate of complexation of metals to $N–CH_3H_2TPPS_4{}^{3-}$ can be used to determine concentrations of copper and zinc in serum.[51]

Acid Promoted Demetalation

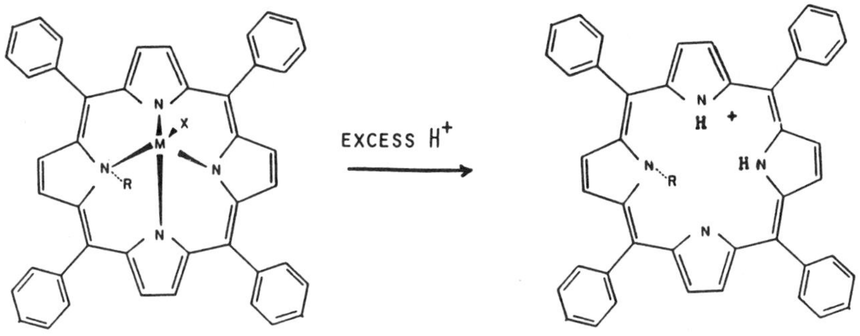

EXCESS H⁺

Figure 4.3. Acidic demetalation of an N−substituted metalloporphyrin.

There are few quantitative data available for acidic demetalation of N−substituted metalloporphyrins, the sole report being the study of the zinc(II) complex of etiopor-phyrin III by Hambright and coworkers.[2] Among those working with N−substituted metalloporphyrins, however, it is common knowledge that many of these complexes are highly sensitive to acid and readily demetalated at acid concentrations much less than those necessary for demetalation of corresponding non−N−substituted metalloporphyrins. The loss of iron from heme when N−substituted porphyrins are formed in reactions of cytochrome P−450 or of hemoglobin with hydrazines, for example, is well documented (Chapters 7 and 8). Some complexes are evidently reasonably stable, however, since the zinc(II) complexes of N−alkylporphyrins inhibit ferro-chelatase in media at physiological pH differently than the corresponding free bases and their presence in solution is readily verified spectroscopically (Chapter 6).

We have observed the behavior of a series of complexes of N-methyltetrakis(4-N-methylpyridyl)porphyrin is aqueous buffer solutions and have found that the complexes of mercury(II), cadmium(II), iron(II) and manganese(II) are readily demetalated at or near physiological pH but that the complexes of zinc(II), cobalt(II), nickel(II) and copper(II) are significantly more stable – to at least pH 5 for several hours.

Hambright and coworkers studied the acid promoted demetalation of zinc(II) complexes of etioporphyrin III and N-methyletioporphyrin III in methanol acidified with HCl.[2] The rate law for etioporphinatozinc(II) was typical for non-N-substituted metalloporphyrins, being third order in hydrogen ion at low concentration and second order at high concentrations:

$$k_{obsd} = k[H^+]^3/(\rho + [H^+])$$

where $k = 2.0 \times 10^5$ $M^{-2}s^{-1}$ and $\rho = 6.6 \times 10^{-3}$ M at 25^0 C. However, the rate law for N-methyletioporphinatozinc(II) is one order lower in acid concentration:

$$k'_{obsd} = k'[H^+]^2/(\rho' + [H^+])$$

where $k' = 2.5 \times 10^2$ $M^{-1}s^{-1}$ and $\rho' = 0.25$ M at 25^0 C. In both cases, the rate law indicates that cleavage of the last strong metal–nitrogen bond is relatively fast (the fourth in the case of the non–N–substituted complex and the third in the case of the N-methylporphyrin complex). Under the conditions used by Hambright, the rates for the two zinc(II) complexes are the same at an acid concentration of about 3×10^{-5} M and the rate increases more rapidly above that pH for the non–N–substituted porphyrin by virtue of the third order dependence. The second order dependence becomes most important at about 0.01 M, so the changes in rate are then more comparable for the two types of porphyrin. The relative behavior for complexes of other metal ions will be, for the most part, difficult to determine because many non–N–

substituted metalloporphyrins cannot be readily demetalated. It is evident, however, that none of the N-substituted metalloporphyrins made to date are stable in acidic solutions at pH values lower than 3. Kinetic data would be of interest with regard to possible application of N-substituted metalloporphyrins as synthetic intermediates or as medicinal agents (Chapter 1).

Metal ions can also be removed from N-substituted porphyrin complexes under nonacidic conditions (for example, iron(II) is readily removed by thiols and thiolates[20]). This type of reaction has not been studied in detail.

Redox Reactions of N-Substituted Metalloporphyrins

The redox reactions of metalloporphyrins have been intensively studied not only for their intrinsic interest to chemists but because of their relevance to the important role of metalloporphyrins in biological redox processes such as the oxidative phosphorylation sequence involving several proteins which contain iron porphyrins. The formation of N-alkylporphyrins *in vivo* very likely includes a redox reaction as an important step and, therefore, redox properties of the N-substituted metalloporphyrins is of biochemical as well as chemical interest. The first part of this section deals with trends in reduction potentials of a variety of metal complexes of N-substituted porphyrins and the latter part deals specifically with reduction of the N-substituted porphyrin complexes of cobalt and iron to form σ-bound metalloporphyrins.

Cyclic Voltammetry. One of the early observations of the properties of N-alkylporphyrin complexes was the stability in air of lower oxidation states of metal ions (for example, Mn(II) and Co(II) rather than Mn(III) and Co(III)[21] and Fe(II) rather than Fe(III)[22]). Cyclic voltammograms revealed that in the N-methylporphyrin complex there is a significant stabilization of the the divalent oxidation state for both manganese and iron. For Mn(N–CH$_3$TPP)Cl$^+$ in comparison to Mn(TPP)Cl, the reduction potentials for Mn(III) to Mn(II) are 0.78 and –0.21 V, respectively, all potentials given vs. Ag/AgCl[21] and for Fe(N–CH$_3$TPP)Cl$^+$ and Fe(TPP)Cl, 0.51 and –0.27 V, respectively).[22] The stabilization of Co(II) is more modest (0.77 vs. 0.52 V for the N–CH$_3$TPP and TPP complexes, respectively).[21] A remarkable feature of the stabilization of lower oxidation states is the appearance of a peak attributable to production of a Cu(I) complex of N–CH$_3$TPP at –0.38 V (for reduction of Cu(N–CH$_3$TPPCl).[23] Typically, no peak for a Cu(II) to Cu(I) process is typically found for non–N–substituted complexes. These N–methylporphyrin complexes all exhibit highly reversible redox processes, unlike the non–N–substituted metalloporphyrins, which often undergo ligation and sometimes significant geometric changes upon reduction of the metal ion. The only electrochemically non–reversible system we have encountered is that for the chromium complex of N–CH$_3$TPP which undergoes reduction (without decomposition) at about –0.8 V and reoxidation (without decomposition) at about –0.16V. The N–substituted porphyrin complexes of chromium, copper, manganese and zinc, and probably most other metals, show the normal pattern of two oxidation waves attributable to oxidation of the porphyrin ring and two reduction waves attributable to ring reductions, with separations typical of other metalloporphyrins. The cobalt and iron complexes, however, undergo a chemical reaction on reduction below the oxidation states Co(II) and Fe(II) (see discussion below).

The effect of the substituents on the periphery of the porphyrin ring on reduction potentials has been determined in a few cases by comparison of potentials of N–substituted tetraphenylporphyrin complexes with N–methylprotoporphyrin IX dimethyl ester, N–methyldeuteroporphyrin IX dimethyl ester, or N–methyloctaethylporphyrin and, at least in these cases, the potentials are influenced modestly by the ring substituents. For example, the Co(III) to Co(II) potentials for chloro complexes of N–CH$_3$TPP and deuteroporphyrin IX dimethyl ester are 0.77 and 0.72 V,[21], those for the Fe(III) to Fe(II) process in the chloro complexes of N-methyltetraphenylporphyrin and N-methylprotoporphyrin IX dimethyl ester are 0.50[22] and 0.39[20] and those for the Fe(III) to Fe(II) process in the complexes of N-phenyltetraphenylporphyrin and N-phenyloctaethylporphyrin are -0.06 and -0.18 volts vs. S.C.E.[24] Certainly, no generalizations about the effect of ring substituents on the redox properties of the metal atom in N-substituted metalloporphyrins are justified.

The effect of the N-substituent on the reduction potential of the bound metal atom and, to a more limited extent, the effect of the nature of the axial ligand, have been determined for a series of N-substituted tetraphenylporphyrin complexes.[23] As shown in Table 4.3, the differences in potentials for a series of complexes of a particular metal atom are modest (on the order of tens of millivolts), whereas the effect of changing the axial ligand can be much greater. The reduction potentials for the Cu(II) to Cu(I) process show that a relatively hard ligand, chloride, gives rise to the least favorable reduction while a poorly coordinating ligand, perchlorate, allows easier reduction and a soft ligand that could be expected to stabilize the cuprous state, triphenylphosphine, gives rise to the most favorable reduction of all the ligands studied. The visible absorption spectra of the copper(II) complexes with chloride and triphenylphosphine as axial ligands are essentially the same and the lack of any shift in potentials on addition of excess ligand shows that the stoichiometry of both complexes is 1:1.

The change in reduction potential as a function of the nature of the axial ligand is that expected if the redox process being monitored is principally centered on the metal ion. Although solvent effects can be substantial for metalloporphyrins and are of interest with respect to the interpretation of differences in the potentials of metalloprotein electron carriers which have iron porphyrins at the active site, no systematic comparisons of solvent effects have been carried out for N-substituted metalloporphyrins.

Table 4.3 Half-Wave Potentials of N-Substituted Tetraphenyl-
porphinato Complexes of Copper, Iron and Manganese[a]

Copper(II) Complexes

Complex	Cu(II)/Cu(I) $E_{1/2}$	ligand Ox$_1$	Ox$_2$	Red$_1$	Red$_2$	$\Delta E_{1/2}$ Ox-Red
Cu(N-CH$_3$TPP)ClO$_4$	-0.29	1.29	1.53	-1.15	-1.36	2.44
Cu(N-CH$_3$TPP)Cl	-0.38			-1.51	-1.70	
[Cu(N-CH$_3$TPP)-(PPh$_3$)]ClO$_4$	-0.04					
Cu(N-C$_2$H$_5$TPP)ClO$_4$	-0.25	1.31	1.52	-1.14	-1.41	2.45
Cu(N-C$_2$H$_5$TPP)Cl	-0.38			-1.28	-1.50	
Cu(N-PhTPP)ClO$_4$	-0.32	1.24	1.46	-1.01	-1.27	2.25
Cu(N-PhTPP)Cl	-0.42				-1.46	

Iron(II) Complexes[b]

Complex	Fe(III)/Fe(II)	Ox$_1$	Red$_1$	Ox-Red
Fe(N-CH$_3$TPP)$^+$ [c]	-0.06		-0.97	
Fe(N-CH$_3$TPP)Cl	0.50	1.52	-0.85	2.37
Fe(N-C$_2$H$_5$TPP)Cl	0.51		-0.86	
Fe(N-PhTPP)Cl	0.54		-0.83	

Manganese Complexes

Complex	Mn(III)/Mn(II) $E_{1/2}$	ligand Ox$_1$	Ox$_2$	Red$_1$	Red$_2$	$\Delta E_{1/2}$ Ox-Red
Mn(N-CH$_3$TPP)Cl	0.82	1.19	1.40	-1.10	-1.31	2.29
Mn(N-PhTPP)Cl	0.82	1.12	1.32	-1.00	-1.28	2.12
Mn(N-p-CH$_2$C$_6$H$_4$-NO$_2$TPP)Cl	0.91	1.24		-0.85	-1.14	2.09

[a] Ref. 23 unless specified. Solvent CH$_3$CN except the
iron complexes (DMF for all but Fe(N-CH$_3$TPP)Cl), which is
CH$_2$Cl$_2$) and Mn(N-CH$_3$TPP)Cl (PhCN/CH$_3$CN), electrolyte 0.1M
TBAP or TEAP. Under these conditions, ΔE_{pa-pc} for ferro-
cenium ion/ferrocene = 0.075 $\pm$ 0.005 V and for these complexes, 0.075 to 0.090 V.

[b] Potentials vs. S.C.E. The oxidation and reduction may be
predominantly metal processes. The reduction only appears
reversible at rapid scan rates.

[c] Ref. 24.

Migration of the N-Substituent on Reduction of Iron(II) and Cobalt(II) N-Substituted Porphyrin Complexes. In 1974, Ogoshi and coworkers reported that reduction of N-methyloctaethylporphinatocobalt(II) with sodium borohydride produces the σ-bound product, methylcobalt(III) octaethylporphine.[25] Dolphin and coworkers carried out the analogous reaction with Co(N-ethylTPP)[+] to obtain ethylcobalt(III)TPP[+].[26] The nature of this type of reversible redox-migration reaction is illustrated in Figure 4.4 Callot and coworkers reported that the reduction of the complex Co(N-PhTPP)OAc with sodium borohydride produces phenylcobalt(III) TPP in high yield (83%) and the analogous reaction with the N-α-styryl complex proceeds in 88% yield. They used deuterated porphyrin ligands and mass spectrometry to deduce that the migration reaction is intramolecular.[27,28] They also reinvestigated the reaction of Co(N-C_2H_5TPP)OAc to form ethylcobalt(III) TPP[+] and found it to be intramolecular as well.[28] Johnson and coworkers demonstrated that the reductive migration reaction proceeds with N-methyloctaethylporphyrin[9] and Ogoshi and coworkers carried out this reaction with the N-methyloctaethylporphyrin complexes of iridium[29] and rhodium.[30,31] Apparently, these reactions occur because the N-substituted cobalt(I), iridium(I) or rhodium(I) can be converted to relatively stable σ-bound cobalt(III), iridium(III) and rhodium(III) complexes by a formal two-electron transfer of the electrons of the carbon-nitrogen bond to form a fourth strong metal-nitrogen bond and, in the case of these metals, formation of a metal-carbon bond (which occurs simultaneously) is relatively favorable.

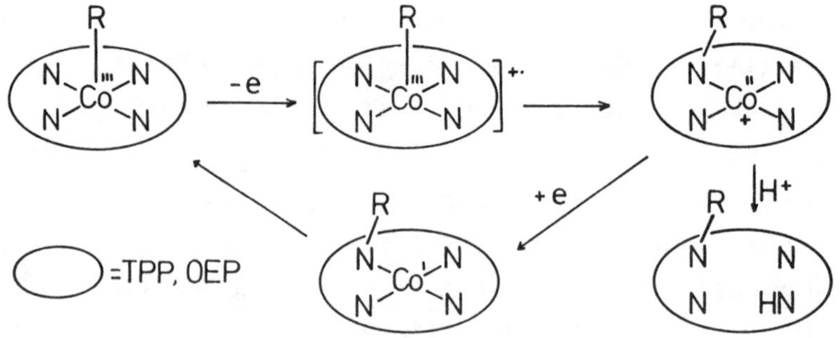

Figure 4.4. The oxidative metal–to–nitrogen and reductive nitrogen–to–metal migration reactions of cobalt porphyrins.

The σ–bound complexes of iron porphyrins are also relatively stable and, since reactions of iron complexes are of relevance to biological processes, several groups have investigated their reactions. Mansuy and coworkers demonstrated that nitrogen to iron migration occurs for the vinylic group $=CH=C(C_6H_5)_2$ upon reduction of the iron(II) complex with sodium dithionite.[32] Kadish and Guilard and coworkers demonstrated that an N–phenyl group in the iron complex of either OEP or TPP migrates upon electrochemical reduction to quantitatively form the σ–phenyliron(III) porphyrin complex.[24] The migration of a group from the nitrogen to metal atom of iron complexes is reversible and the reaction in the opposite sense – oxidative migration of an aryl group from the iron atom to the nitrogen atom that – has received the most attention (Chapter 8). Brothers and Collman and Setsune and Dolphin have discussed these reactions in the context of the general organometallic chemistry of metalloporphyrins in recent reviews.[50,52]

Nucleophilic Displacement of the N-Substituent

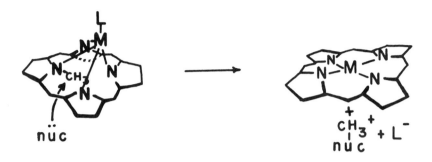

Figure 4.5. Nucleophilic displacement of an N-substituent.

Background. Shears and Hambright reported in 1971 that the zinc(II) complex of N-methyletioporphyrin III lost the N-methyl group under conditions (refluxing for 3 days in pyridine) where the free ligand did not lose its N-substituent. While attempting to measure the metalation of N-CH$_3$HTPP by metal salts in dimethylformamide,[11] we noticed that a second reaction occurs on the same time scale as metalation in the cases of copper(II) and nickel(II), and we found that this reaction was demethylation. The facility of these reactions, in contrast to the rather severe conditions needed to demethylate the zinc(II) complex and even harsher conditions needed for free bases, encouraged us to investigate this type of reaction in detail. After establishing the rate law and the general nature of the reaction, we investigated effects of the nature of the metal ion, the substituents on the porphyrin ring periphery, the solvent, the nucleophile and the N-substituent. Under proper conditions, the removal of the N-substituent occurs so rapidly that the two step reaction of metalation of the N-substituted porphyrin and removal of the substituent can occur much more rapidly than the direct formation of a complex

between the metal ion and the corresponding non-N-substituted porphyrin. Rapid porphyrin metalation, especially in
aqueous solvents at neutral pH, is of interest for the
formation of complexes of radioactive metal ions with short
half-lives that can be of medical use.

General Aspects of Dealkylation Reactions. Initial
investigations of nucleophilic displacement reactions of N-
methyltetraphenylporphyrin demonstrated that the reactions
were highly sensitive to the nature of the nucleophile, with
aliphatic amines acting as much stronger nucleophiles than
pyridine and oxygen donors such as alcohols unreactive.[33]
Reactions were found to proceed at convenient rates in the
solvent acetonitrile which is stable and readily purified.
Thus, most results for this type of reaction have been
derived using the relatively non-volatile and stable alkyl
amine, di-n-butylamine, in acetonitrile. The one major
drawback to this system is that acetonitrile is not as good
a ligand for some metals as the di-n-butylamine, so that
while most metals give a rate law that is simply first order
in di-n-butylamine concentration, the reactions of the
copper(II) and nickel(II) N-substituted porphyrins have a
rate law that is more complex:

$$-d[MN\text{-}RP]/[MN\text{-}RP]dT \;=\; (k_1[nu] + k_2 K_{eq}[nu]^2)/(1 + K_{eq}[nu])$$

where we have interpreted the parameters in the rate law to
represent a path involving the solvent as the axial ligand
(k_1), a path with the nucleophile, di-n-butylamine, as the
axial ligand (k_2) and an equilibrium constant for the interconversion of the two forms (K_{eq}). Although there are other
kinetically equivalent mechanisms to arrive at this form of
a rate law, the activation parameters (and even the absolute
magnitudes of the rate constants) for the two paths are
similar so comparisons should still be meaningful. Also,
the magnitude of the calculated K_{eq} values are reasonable
for complexes of Cu(II) and Ni(II) in acetonitrile.[33,34]
Other metal ions having rate laws first order in the nucleo-

phile are apparently all in one form or the other under the
conditions studied. The rate laws for reactions in other
media will be discussed below.

The Effect of the Metal Atom. Under the same conditions,
the rates for demethylation of a variety of N-substituted
metalloporphyrins (in the temperature range of $15^\circ - 79^\circ C$)
has been found to be in the order Pd(II) > Cu(II) > Ni(II) >
Co(II) > Zn(II) > Mn(II). This sequence corresponds to the
stability of the non-N-substituted porphyrin product, with
the more stable (and often more planar) products being
generated more rapidly. Activation parameters have been
determined for most of these metal ions (Table 4.4) and they
show more favorable enthalpies and free energies for the
reactions that proceed more rapidly near ambient temperature. We interpret these results to indicate that the
activated complex strongly resembles the product (that there
is a significant degree of SN_1 character to the reaction).
The rate law indicates that the nucleophile is present in
the activated complex. The sensitivity of the rate to the
nature of the nucleophile indicates that bond formation is
also important.

Table 4.4 Activation Parameters for the Demethylation of
N-Methyltetraphenylporphyrin Complexes by Di-n-Butylamine in
Acetonitrile[a]

Metal Ion	$\Delta H^{\ddagger}$ (kcal/mol)	$\Delta S^{\ddagger}$ (eu)	$\Delta G^{\ddagger}$ (318 K) (kcal/mol)
Pd(II)	14.1 ± 0.6	−21.5 ± 1.6	21
Cu(II)	16.9 ± 1.0	−13.1 ± 2.9	21
Ni(II)	18.0 ± 1.0	−14.0 ± 3.0	23
Zn(II)	41.6 ± 2.0	−54.6 ± 6.2	26
Mn(II)	> 51		

a Refs. 34, 35.

Effect of the Porphyrin Ring Substituents. The N-alkyl-porphyrins formed in nature are all derivatives of protoporphyrin IX while many studies of porphyrins are carried out on synthetic *meso*-substituted porphyrins. Hence, it is of interest to compare the reactivities of the two classes of N-substituted porphyrins. In addition, applications involving N-substituted metalloporphyrins can be facilitated when the effects of structural modifications on reactivity can be predicted. Therefore, we carried out reactions under the same conditions for nucleophilic displacement of three very different porphyrins. As is evident from Table 4.5, these reactions proceed at nearly the same rate and with very similar activation parameters despite the variation in peripheral substituents from four neutral phenyl groups (N-CH$_3$TPP) to four anionic p-sulfonatophenyl groups (N-CH$_3$TPPS$_4$$^{4-}$) to the non-*meso*-substituted deuterporphyrin IX dimethyl ester (N-CH$_3$DP). Hence, it is possible to compare the results for the nucleophilic displacement reaction of one metalloporphyrin with another if the only difference in the reaction conditions is the identity of the peripheral substituents.

Table 4.5. Rate Constants and Activation Parameters for the Demethylation of Several Copper(II) N-Methylporphyrins[a]

Complex	k_{obsd}[b] (s^{-1}, x 10^3)	$\Delta H^{\ddagger}$ (kcal/mol)	$\Delta S^{\ddagger}$ (eu)	$\Delta G^{\ddagger}$ [c] (kcal/mol)
Cu(N-CH$_3$TPP)$^+$	1.3	16.9	−13.1	20.8
Cu(N-CH$_3$TPPS$_4$)$^{3-}$	1.6	15.6	−17.6	20.8
Cu(N-CH$_3$DP)	0.6	15.3	−20.2	21.3

[a] Ref. 36 for reactions in acetonitrile with di-n-butylamine as the nucleophile.
b. At 45° C, with [di-n-butylamine] = 0.10 M, where the path with solvent as the axial ligand (k$_1$ path) predominates.
c. At 298 K.

The Effect of the Reaction Medium. In acetonitrile, dealkylation of $Cu(N-CH_3TPP)^+$ by chloride ion is very slow. However, Stinson and Hambright found that $Cu(N-CH_3TPP)Cl$ tranforms spontaneously into CuTPP when it is dissolved in chloroform, with a half life of 75 min. The reaction rate is independent of the concentration of the complex ($\Delta H^{\ddagger}$ = 24.4 $\pm$ 0.5 kcal/mol, $\Delta S^{\ddagger}$ = 3.7 $\pm$ 1.0 eu) with methylchloride as a product.[37] Thus, in chloroform, the chloride ion acts as a much more powerful nucleophile than it does in acetonitrile. This reaction also proceeds in other solvents of low polarity such as toluene, dichloromethane and carbon tetrachloride, but is orders of magnitude slower in solvents that can solvate ions, such as methanol or acetonitrile. Using $Cu(N-CH_3TPP)CF_3SO_3$ with a range in concentrations of tetraethylammonium chloride in dichloromethane, we found that the reaction is independent of chloride ion as long as there is more chloride ion than porphyrin complex.[36] The independence of the rate constant on concentration implies that the reaction arises from the intramolecular reaction of a tight ion pair. It also implies that the chloride ion can be found on the correct side of the porphyrin to attack the methyl group which is surprising because that is the side opposite the copper(II) atom.

Because of the independence of the demethylation rate on chloride concentration in dichloromethane, but the marked dependence on di-n-butylamine concentration, we conclude that the chloride ion is a stronger nucleophile when both nucleophiles are at low concentration. The better the solvent solvates ions, the more that nitrogen nucleophiles are favored over chloride ion. As expected from the behavior typical of other nucleophilic displacement reactions, reactions with the same nucleophile are retarded when the polarity of the reaction medium is increased. For example, the reaction of $Cu(N-CH_3TPPS_4)^{3-}$ with di-n-butylamine is orders of magnitude slower in water than in acetonitrile.[36] The effect of solvent has been demonstrated for the reaction of palladium(II) with $N-CH_3HTPP$ and di-n-butylamine in acetonitrile, dimethylsulfoxide and dimethylformamide.[35] In acetonitrile, there are clear isosbestic points early in the reaction showing rapid formation of the complex Pd(N-CH$_3$TPP)$^+$ which is demethylated in the rate determining step. In dimethylformamide, the spectra show isosbestic points for the free base ligand and the product PdTPP, showing that the rate determining step is formation of Pd(N-CH$_3$TPP)$^+$. In dimethylsulfoxide, the behavior is not so clear, but it appears that an outer-sphere complex with a distinct spectrum may be formed rapidly. Thus, the nature of the mechanism, the order of reactivity of nucleophiles and the rate for a specific nucleophile are all profoundly affected by the nature of the solvent medium in which the reaction is carried out. The reactions of N-methylporphyrin complexes in aqueous solution are too slow to be of interest for rapid synthesis of non-N-substituted compounds.

Effect of the N-Substituent. The availability of a variety of methods to produce a wide range of N-substituted porphyrins has allowed us to investigate the effect of the type of N-substituent on the rate of nucleophilic displacement (see Chapter 5 for a description of synthetic methods). Since the earlier kinetics studies had demonstrated a high degree of SN$_1$ character to the reaction, we decided to investigate N-substituents which had a good carbocation stabilizing substituent: a benzyl group. At the time of the study, the only published synthesis was limited to the formation of N-(p-nitrobenzyl)HTPP, but we have now developed an alternative synthesis and studied other benzyl groups with similar results. We also investigated reactions in which the N-substituent had an sp^2 carbon atom bound to the pyroleninic nitrogen atom (N-phenylHTPP) and those with alkyl (N-CH$_3$HTPP and N-CH$_2$H$_5$HTPP) and acyl (N-ethyl-acetatoHTPP) groups. The N-phenylporphyrins are very unreactive, their Cu(II) complexes remaining intact for weeks at 60° C in acetonitrile with 1 M di-n-butylamine. In the same medium, the rate for loss of the N-p-nitrobenzyl group is about 90 times faster than the N-methyl group. Loss of an N-ethyl group is about 10 times slower than the

N-methyl group and the reaction involving N-ethylacetatopor-
phyrin is intermediate between the two. The activation
parameters for these reactions were obtained using acetonitrile as solvent and di-n-butylamine as the nucleophile.
These parameters (Table 4.6) indicate that the principal
advantage of the more reactive groups is enthalpic, consistent with the high degree of SN_1 character of the reaction
and the degree to which these ligands can stabilize charge
development in the transition state. These reactions have
similar rates for *meso*-substituted porphyrins such as tetraphenylporphyrin, tetrakis(p-sulfonatophenyl)porphyrin, and
tetrakis(p-carboxyphenylporphyrin, as well as and non-*meso*-
substituted porphyrins such as protoporphyrin, deuteroporphyrin and hematoporphyrin.[38] It is, therefore, quite
clear that the nature of the N-substitutent has a much
greater effect on dealkylation rates than the peripheral
substituents. For a particular metal ion, the best conditions of dealkylation are a solvent of low polarity, a good
nucleophile and an N-substituent which can accommodate positive charge in the activated complex.

Table 4.6. Rate Constants and Activation Parameters for the
Nucleophilic Displacement of N-substituents From Copper(II)
Porphyrin Complexes by Di-n-butylamine in Acetonitrile.[a]

Complex	k_{obsd}[b] $(s^{-1}, \times 10^3)$	$\Delta H^{\ddagger}$ (kcal/mol)	$\Delta S^{\ddagger}$ (eu)	$\Delta G^{\ddagger}$ [b] (kcal/mol)
$Cu(N-CH_3TPP)^+$	2.9	16.9	-13.1	20.8
$Cu(N-C_2H_5TPP)^+$	0.23	18.2	-13.9	22.4
$Cu(N-p-CH_2C_6H_4-NO_2TPP)^+$	271	14.8	-11.6	18.3

a   Ref. 40, activation parameters are given for the
path at low nucleophile concentration (the k_1 path).
b   At 25° C, the low concentration path.

Even more striking effects of the nature of the N-substituent on the delalkylation rate are found in aqueous solutions and alcoholic media. There the N-benzyl complexes of palladium(II), copper(II) and cobalt(II) all readily react with the solvent acting as nucleophile to form the non-N-substituted porphyrin. Under the same conditions, N-methyl-porphyrin complexes are almost unreactive. Using the N-benzyl substituent, non-N-substituted porphyrins of palladium(II), copper(II) and cobalt(II) can be synthesized under conditions where proteins are stable (neutral aqueous buffers). Under the following conditions: 0.1 M borate buffer at pH 7.5, 1.0×10^{-5} M metal ion, 1.0×10^{-6} M porphyrin, the half-lives for formation of the non-N-alkyl metalloporphyrin from N-p-nitrobenzyltetrakis(p-sulfonato-phenyl)porphyrin at 40° C are 1 minute for palladium(II) (from tetrachloropalladate), 2 minutes for cobalt(II) (from cobaltous acetate) and 4 minutes for copper(II) (from cupric chloride) (unpublished results).   This method has been applied to the synthesis of radiolabelled monoclonal antibodies for medical diagnosis.[39] For this application, it is necessary to have a functional group available on the periphery of the porphyrin that can from a covalent link with the protein. We chose the carboxylate group for our initial studies. The carboxylate (of either N-benzyltetrakis(p-carboxyphenylporphine, the simplest to synthesize or of 5-(p-carboxylphenyl)-10,15,20-tris(p-sulfonatophenyl)-porphine which has the advantage of only as single coupling functionality)is linked to the protein at lysine residues using a carbodiimide coupling agent. The protein-porphyrin conjugate is then purified before the metal isotope is added.   The metalation and dealkylation reaction is essentially quantitative, so no separation processes are required after the radioactive metal salt has been added. An important feature of the mechanism is that metal complexation (in the cases of copper(II) and cobalt(II) is very rapid compared to dealkylation, so that the overall reaction rate is independent of the concentration of the radioactive metal ion.

Formation of Homoporphyrins From N-Substituted Porphyrins

Background. Callot and coworkers have found that zinc(II) and nickel(II) N-substituted porphyrins can give rise to products that result from the transfer of the substituent to the periphery of the porphyrin.[41,42,43] The products of these reactions possess one two-carbon *meso* bridge and are called "homoporphyrins." Callot and coworkers studied the reaction of N-ethylcarbonylmethylenetetraphenylporphyrin with nickel(II) salts especially thoroughly.[42,44-46] The overall reaction is depicted in Figure 4.6. Using mild reaction conditions, they were able to isolate the intermediates that are shown in Figure 4.7.[42] The process appears to occur by sequential migration of an alkyl group from the metal atom to a nitrogen atom, followed by loss of a proton and formation of a unique type of porphyrin that has an aziridine ring bridging the nitrogen and an α-carbon of a pyrrolenine ring.

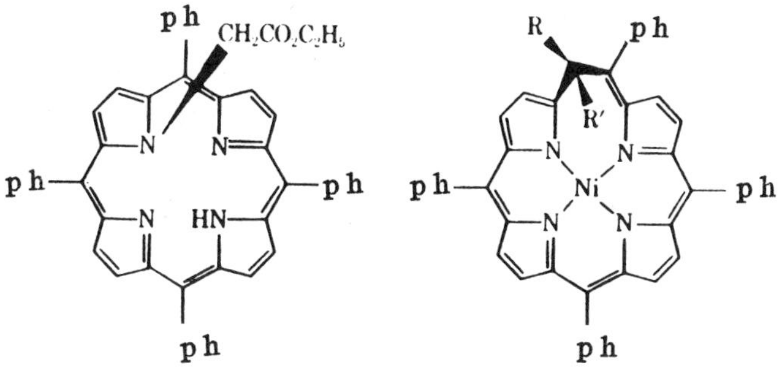

Figure 4.6 The N-ethyoxycarbonylmethylenetetraphenylporphyrin (left) which reacts with Ni(II) to produce homoporphyrins (right). The R and R' groups are H and $CO_2C_2H_5$. When R = H, the isomer is endo (the predominate product) and when R' = H, the isomer is exo.[42]

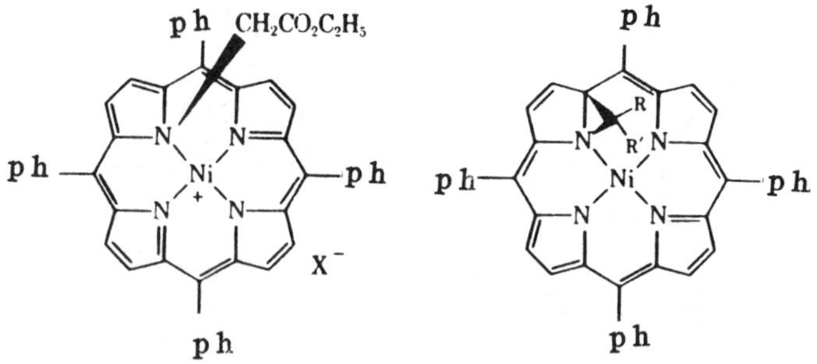

Figure 4.7 Intermediates in the formation of homoporphyrins. The first intermediate is chloro-N-ethylcarbonylmethylene-tetraphenylporphinatonickel(II) (left) and the second is a unique aziridine porphyrin (right).[42]

Reactions of Nickel Complexes. The first intermediate, the chloronickel(II) complex of N-ethylcarbonylmethyleneTPP, was obtained in 75% yield by refluxing N-ethylcarbonylmethyleneHTPP with nickel(II) acetate in refluxing chloroform/methanol (1:1) and adding aqueous sodium chloride to precipitate the chloride salt. Dissolution of this intermediate in triethylamine and dichloromethane (1:1 v/v) and stirring for 33 h at room temperature produced a second intermediate, an unusual aziridine, isolated in 45% yield. When the aziridine derivative was refluxed in benzene, the product (as the endo isomer) was obtained in quantitative yield. The overall reaction of the initial free base porphyrin with a mixture of nickel(II) acetate and nickel(II) carbonate in refluxing 1,2-dichloroethane gives a mixture of the endo and exo isomers (in 57% and 4% isolated yields, respectively).[42] On heating, the endo and exo isomers interconvert.

The analogous reaction with N-ethylcarbonylmethyleneoctaethylporphyrin leads to a *meso*-substituted product (in 39%

yield) rather than a homoporphyrin (Figure 4.8).[42] The authors noted that a green colored intermediate was formed during the reaction which may have been the homoporphyrin, but no intermediates could be isolated. The free base form of the product was obtained using concentrated sulfuric acid to remove the nickel atom and was also obtained, though in low yield, from the reaction of CuOEP with ethyldiazoacetate followed by acidic demetalation.

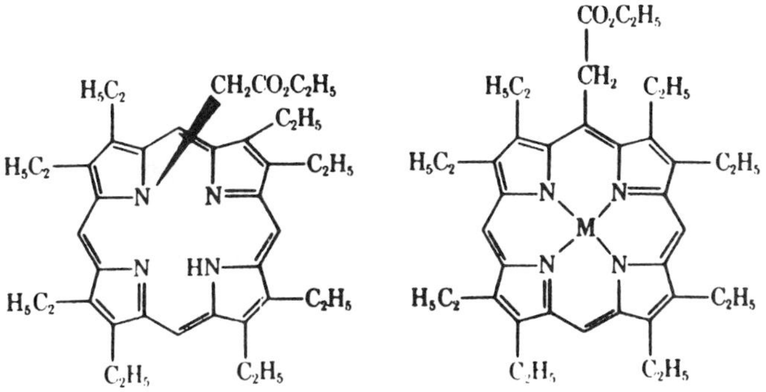

Figure 4.8. The N-ethylcarbonylmethyleneoctaethylporphyrin (left) which reacts to form a *meso*-substituted product (right).[42]

Reactions of Zinc Complexes. Johnson and coworkers found that treatment of ZnTPP with methyldiazopropionate and other diazo derivatives (with the exception of diazoacetate, which gives N-ethoxycarbonylmethyleneHTPP) produced homoporphyrin products.[47] Callot and Schaeffer later demonstrated that both N–CH(CH₃CO₂C₂H₅)HTPP or the homoporphyrin with the two-carbon *meso* fragment C(CH₃)(CO₂C₂H₅)C(C₆H₅) can be isolated from the reaction of methyldiazopropionate and ZnTPP, the ratio depending on the amount of water present in the reaction medium.[43]

The mechanism was proposed to account for this result involves addition of the diazo compound to the zinc(II) atom, loss of nitrogen, and formation of the N-substituted zinc complex which can form an aziridine intermediate (as isolated in the case of the nickel(II) reaction discussed above) that may proceed to form a homoporphyrin or revert to the N-substituted form by protonation. In the presence of protons, the N-substituted complex may be demetalated to form the free base. The zinc(II) atom is more readily removed from N-substituted complexes by acid than is a nickel(II) complex. This mechanism explains the effect of water, which is a proton source. The yield for the homo-porphyrin in the absence of water is 52%, but with 0.05 mL of water in 5 mL of xylene, methyldiazopropionate produces 35% of the N-substituted product and only 13% of the homo-porphyrin.[43] The reaction of the N-substituted zinc(II) complex is consistent with a general mechanism for migration reactions of a wide variety of N-substituted complexes pro-posed by Callot (is discussed in further detail in Chapter 8). The rearrangement of N-substituted nickel(II) and palladium(II) corroles may also be responsible for the methyl addition at a pyrroleninic carbon observed by Johnson and coworkers.[48,49] Aside form these studies, there has been little utilization of N-substituted porphyrins to pro-duce modified porphyrins. There now exists an opportunity for interesting synthetic application of N-substituted pre-cursors because new synthetic methods are available to introduce a wide variety of N-substituents.

Conclusion.

Although few of the reactions of N-substituted porphyrins aside from metalation and nucleophilic displacement of the N-substitutent have been studied in detail, a number of characteristics of the reactions of these compounds have been established. The relatively basic nature of these compounds and the facile acidic demetalation of their complexes are certainly important properties to be kept in mind when handling them. The use of the N-substituent as a reactive moiety has received only scant attention and may lead to interesting new reactions.

References

1. B. Shah, B. Shears and P. Hambright, Inorg. Chem., **10** (1971) 1828-1830.

2. B. Shears, B. Shah and P. Hambright, J. Am. Chem. Soc., (1971) 776-778.

3. B. Shears and P. Hambright, Inorg. Nucl. Chem. Lett., **6** (1970) 679-680.

4. W.K. McEwen, J. Amer. Chem. Soc., **58** (1936) 1124-1129.

5. A. Neuberger and J.J. Scott, Proc. Roy. Soc., **A213** (1952) 307-326.

6. R. Grigg, R.J. Hamilton, M.L. Josefowicz, C.H. Rochester, R.J. Terrell and H. Wickwar, J. Chem. Soc., Perkin II, (1973) 407-413.

7. D.K. Lavallee and A.E. Gebala, Inorg. Chem., **19** (1974) 2004-2008.

8. K.A. Freeman and F. Hibbert, J. Chem. Soc., Perkin II, (1979) 1574-1578.

9. H.M.G. Al-Hazimi, A.H. Jackson, A.W. Johnson and M. Winter, J. Chem. Soc. Perkin I, (1977) 98-103.

10. W.K. McEwen, J. Amer. Chem. Soc., **68** (1946) 711-713.

11. M.J. Bain-Ackerman and D.K. Lavallee, Inorg. Chem., **18** (1979) 3358-3364.

12. D.K. Lavallee, Coor. Chem. Rev., **61** (1985) 55-96.

13. F.R. Longo, E.M. Brown, D.J. Quimby, A.D. Adler and M. Meot-Ner, Ann. N.Y. Acad. Sci., **206** (1973) 420-442.

14. P. Hambright and P.B. Chock, J. Am. Chem. Soc., **96** (1974) 3123-3127.

15. S. Funahashi, Y. Yamaguchi and M. Tanaka, Bull. Chem. Soc. Jpn., **57** (1984) 204-208.

16. D.K. Lavallee and G. Onady, Inorg. Chem., **20** (1981) 907–908.

17. S. Funahashi, Y. Yamaguchi, K. Ishihara and M. Tanaka, J. Chem. Soc., Chem. Commun., (1982) 976–977.

18. S. Funahasi, Y. Yamaguchi and M. Tanaka, Inorg. Chem., **23** (1984) 2249–2251.

19. A. Adeyemo, A. Shamin, P. Hambright and R.F.X. Williams, Ind. J. Chem., **21A** (1982) 763–766.

20. D.K. Lavallee, J. Inorg. Biochem., **16** (1982) 135–143.

21. D.K. Lavallee and M.J. Bain, Inorg. Chem., **15** (1976) 2090–2093.

22. O.P. Anderson, A.B. Kopelove and D.K. Lavallee, Inorg. Chem., **19** (1980) 2101–2107.

23. D. Kuila, A.B. Kopelove and D.K. Lavallee, Inorg. Chem., **24** (1985) 1443–1446.

24. D. Lancon, P. Cocolios, R. Guilard and K.M. Kadish, J. Amer. Chem. Soc., **106** (1984) 4472–4478.

25. H. Ogoshi, E.I. Watanabe, N. Koketzu and Z.I. Yoshida, J. Chem. Soc., Chem. Commun., (1974) 943–944.

26. D. Dolphin, D.J. Halko and E. Johnson, Inorg. Chem., **20** (1981) 4348–4351.

27. H.J. Callot and F. Metz, J. Chem. Soc., Chem. Commun., (1982) 947–948.

28. H.J. Callot and R. Cromer, Nouv. J. Chim., **8** (1984) 765–770.

29. H. Ogoshi, J. Setsune and Z. Yoshida, J. Organomet. Chem., **159** (1978) 317–328.

30. H. Ogoshi, T. Omura and Z.I. Yoshida, J. Chem. Soc., **95** (1973) 1666–1668.

31. H. Ogoshi, J. Setsune, T. Omura and Z. Yoshida, J. Amer. Chem. Soc., 97 (1975) 6461–6466.

32. D. Mansuy, J.-P. Battioni, D. Dupré, E. Sartori and G. Chottard, J. Am. Chem. Soc., **104** (1982) 6159–6161.

33. D.K. Lavallee, Inorg. Chem., **15** (1976) 691–694.

34. D.K. Lavallee, Inorg. Chem., **16** (1977) 955–957.

35. J.D. Doi, C. Compito-Magliozzo and D.K. Lavallee, Inorg. Chem., **23** (1984) 79-84.

36. D. Kuila and D.K. Lavallee, Inorg. Chem., **22** (1983) 1095-1099.

37. C. Stinson and P. Hambright, Inorg. Chem., **15** (1976) 3181-3182.

38. D.K. Lavallee, A. White, A. Diaz, J.-P. Battioni and D. Mansuy, Tetrahedron Lett., **27** (1986) 3521-3524.

39. J. Mercer-Smith, S. Figard, D.K. Lavallee and Z. Svitra, J. Nucl. Med.,**26** (1985) 437 and in Antibody-Mediated Delivery Systems, Marcel-Dekker, New York, 1987.

40. D.K. Lavallee and D. Kuila, Inorg. Chem., **23** (1984) 3987-3992.

41. H.J. Callot, Bull. Soc. Chim. Fr., **11** (1972) 4387-4391.

42. H.J. Callot and Th. Tschamber, J. Amer. Chem. Soc., **97** (1975) 6175-6178.

43. H.J. Callot and E. Schaeffer, Nouv. J. Chim., **4** (1980) 307-309.

44. H.J. Callot and Th. Tschamber, Tetrahedron Lett., **36** (1974) 3155-3158.

45. H.J. Callot and Th. Tschamber, Tetrahedron Lett., **36** (1974) 3159-3162.

46. H.J. Callot, Th. Tschamber and E. Schaeffer, J. Amer. Chem. Soc., **97** (1975) 6178-6180.

47. P. Batten, A.L. Hamilton, A. W. Johnson, M. Mahendram, D. Ward and T.J. King, J. Chem. Soc., Perkin I, (1977) 1623-1628.

48. R. Grigg, A.W. Johnson and G. Shelton, J. Chem. Soc., (1971) 2287-2294.

49. R. Grigg, A.W. Johnson, and G. Shelton, Liebigs Ann. Chem., **746** (1971) 32-53.

50. P.J. Brothers and J.P. Collman, Accts. Chem. Res., **19** (1986) 209-215.

51. S. Funahashi, Y. Ito, H. Kakito, M. Imano, Y. Hamada and M. Tanaka, Mikrochim. Acta., **4** (1986) 33-47.

52. J.I. Setsune and D. Dolphin, Can. J. Chem., **65** (1987) 459-467.

CHAPTER 5

CHEMISTRY OF N-SUBSTITUTED PORPHYRINS IV:
SYNTHETIC METHODS

Introduction

The first synthesis of an N-substituted porphyrin, that
of N-methyletioporphyrin I by McEwen in 1936, was accomplished by the direct alkylation of the porphyrin free base
with methyl iodide in a sealed tube.[1] Direct alkylation
reactions with methyl iodide were used subsequently by Jackson and coworkers to produce mono-, di- and trimethyl etioporphyrins and octaethylporphyrins.[2] The more powerful alkylating agents methylfluorosulfonate or methyltrifluorometh-
anesulfonates were then used to produce N-methylated tetraphenylporphyrins.[3],[4] Although direct alkylations can be
simple and convenient, this method is limited by the
availability of appropriate precursors to a few simple alkyl
groups. In the past several years, a number of other
methods for the synthesis of N-substituted porphyrins have
been developed, extending the range of possible compounds
enormously. This chapter will provide preparative methods
for six classes of porphyrins with the following types of N-
substituents 1) one or more alkyl or an alcoxycarbonylmethylene, 2) phenyl or vinyl, 3) benzyl or allyl, 4) one or two
atom moieties with carbon or nitrogen bridging two
neighboring pyrroleneine nitrogens of free base porphyrins,
4) N,N'-(1,2-vinylidene) or N,N'-phenylene moieties which
have a bridge between two pyrrolenine nitrogens consisting
of two carbon atoms, and 6) moieties with one carbon or
nitrogen atom bridging a pyrrolenine nitrogen and the
coordinated metal. In addition, the synthesis of an N-
aminoporphyrin, a total specific synthesis of each of the
four N-methyl isomers of protoporphyrin IX, and the
synthesis of metal complexes will be discussed.

General Comments on Synthetic Methods

Although the syntheses of N-substituted porphyrins are
often well-described in the literature, the tendency to
publish in highly abbreviated form sometimes necessitates
omission of procedural details that may be critical to
successful repetition of a synthesis. A brief description of
some of the common properties and handling procedures of N-
substituted porphyrins may be useful in this regard and in
anticipating difficulties in the synthesis of new N-substituted porphyrins.

The N-substituted porphyrins are typically more polar
than the unsubstituted free base and can often be separated
on this basis. To exploit this difference, of course, the
acid functionalities of such porphyrins as protoporphyrin,
deuteroporphyrin, coproporphyrin or synthetic porphyrins
such as 5,10,15,20-tetrakis(p-carboxyphenyl)porphyrin must
be in the acid or ester rather than the carboxylate form.
In addition to their higher polarity, the N-substituted
porphyrins are generally stronger bases than the corresponding non-N-substituted porphyrin. In many cases, they can
be readily be separated on acidic or even "neutral"
chromatographic supports because they will be monocationic
while the non-N-substituted precursor is neutral. The di-N-
substituted porphyrins are even stronger bases than the
mono-N-substituted porphyrins. Thus, they are generally
bound firmly to alumina or silica and are eluted by deactivating the column with methanolic or aqueous solutions or by
the addition of a base to the eluent mixture. TLC
(generally neutral alumina for *meso*-tetraarylporphyrins and
silica for non-*meso*-tetraarylporphyrins) is typically an
excellent method of monitoring the progress of alkylation
reactions and visible spectroscopy is often useful
(especially in cases where the reaction medium is very
polar).

The stability of these porphyrins with respect to loss of the N-substituent varies widely: The N-phenylporphyrins that can often be heated to $150^{\circ}C$ for days without decomposition; the N-allylporphyrins revert to the corresponding non-N-substituted porphyrin in the solid state at room temperature, and the N,N'-bridged porphyrins are often decomposed by even dilute solutions of base. The discussion in the chapter on the reactions of these porphyrins should be useful in anticipating the reactivity of a new species. In general, it is wise to avoid high temperatures by using a rotary evaporator for removal of solvents and using air drying or, if necessary, a vacuum oven at mild temperatures for the final drying procedure. Contamination by metal ions and prolonged contact with strong nucleophiles such as ammonia should be avoided. (Relatively stable N-substituted porphyrins such as the N-alkyl-, N-phenyl-, N-vinyl, and N-alcoxycarbonylmethyleneporphyrins, however, are often neutralized to their free base forms for separation or crystallization with ammonia but aqueous sodium hydrogen carbonate is appropriate for more reactive species such as the N-benzyl, N-allyl and N,N'-bridged porphyrins.) More highly crystalline products are typically obtained for free bases than the cationic salts.

The basic nature of the N-substituted porphyrins often requires that a competing base, such as pyridine, must be added to obtain a clean NMR spectrum of the free base. Alternatively, the free base can be readily converted in most cases to the zinc(II) complex for NMR spectroscopy. In some cases, use of the zinc complex may provide a means for cleaner chromatographic separation as well. The zinc atom can readily be introduced using zinc acetate in a polar organic solvent such as methanol or acetonitrile and can be removed after analysis or separation using a mild acidic solution and aqueous extraction.

Synthesis of N-Alkyl- and N-Alcoxycarbonylmethyleneporphyrins

Alkylfluorosulfonates and Trifluoromethanesulfonate. The highly reactive methyl- and ethylfluorosulfonates provide a straightforward method for the synthesis of N-methyl- and N-ethylporphyrins of both *meso*-substituted porphyrins such as N-CH$_3$HTPP) and non-*meso*-substituted porphyrins such as N-CH$_3$HOEP. In either case, use of the alkylating agent in a stoichoiometric ratio with the porphyrin leads to predominant formation of the mono-N-substituted product while an excess leads to di- and/or tri-N-substituted products. The mono-N-alkylated porphyrins react more readily with alkylating agents than do non-N-alkylated precursors. Hence, one might expect to have difficulty stopping the reaction after the first alkylation. However, because the mono-N-alkylporphyrins are stronger bases than the precursor porphyrins, these porphyrins are retain both or their original protons in solvents of low basicity, producing a monoprotonated N-alkylporphyrin as the major product. This species is less reactive toward further alkylation than is the free base of the precursor porphyrin, suppressing further alkylation. The selectivity of the mono-N-alkylporphyrin to bind the proton of the free base precursor is much greater for non-*meso*-tetraarylporphyrins than for porphyrins such as H$_2$TPP because the former are much stronger bases.[3],[5] It is, therefore, easier to achieve high yields of the mono-N-substituted forms of those porphyrins. For the *meso*-tetra-arylporphyrins, high dilution conditions favor formation of the mono-N-alkylated product. Addition of a mild base (which does not react with the alkylating agent) to allow the mono-N-alkylated porphyrin to remain in the free base form leads to di- and trialkylated products.

Procedure.[3] The precursor porphyrin is added to dichloromethane at less than 10^{-2} M (4.0 g of H$_2$TPP to 1.5L of CH$_2$Cl$_2$) in a three-neck flask with magnetic stir bar, condenser and 500 mL pressure-equalizing addition funnel. The mixture is brought to reflux and a stoichiometric amount of the fluorosulfonate or trifluoromethane sulfonate diluted in CH_2Cl_2 (for example, 0.74 mL of methyltrifluoromethane-sulfonate in 300 mL of CH_2Cl_2) is added drop-by-drop over a period of 2-4 h. The mixture is refluxed with stirring overnight, allowed to come to room temperature, and reduced in volume to about 750 mL using a rotary evaporator. The bright forest-green solution is then neutralized using about 200 mL of 1M aqueous ammonia. The organic layer is separated and washed with 3 or 4 portions of water, about 150-200 mL each, to remove the ammonia and salt. It is then dried with magnesium sulfate (about 25 g), filtered, and reduced in volume to about 350 ml and filtered again. At this point, the volume should be as small as possible with all of the product still in solution. The product is divided into two equal portions and each is chromatographed on a column of neutral alumina 9 cm in diameter and 50 cm high using a CH_2Cl_2 slurry to prepare the column. The H_2TPP starting material is removed as a burgundy band using CH_2Cl_2 as eluent. The $N-CH_3HTPP$ band is second. It is forest-green on the column and purple as it elutes. It can be eluted with CH_2Cl_2 also or, more rapidly, using 1-5% of a more polar solvent such as acetonitrile. The next green band consists of the di-N-methylporphyrin (probably *cis* and *trans* isomers, but we have not analyzed them carefully). It can be removed with 10% CH_3CN in CH_2Cl_2, if desired. Typical yields of mono-N-methyl-*meso*-tetraarylporphyrins are 30-35%. About 5-8% of the product is di-N-methylporphyrin (as mono-protonated salts) and the remainder is recyclable unmethylated porphyrin. The mono-N-methylated porphyrins can be obtained in high purity by crystallization from CH_2Cl_2/CH_3CN (the porphyrin being dissolved in about 100 mL of $CH_2Cl_2/$ g, followed by filtration and addition of an equal volume of CH_3CN and evaporation at room temperature over a period of a few days to leave about 50 mL of solution). Small amounts impurities may be removed by dissolving the porphyrin in CH_2Cl_2, filtration and precipitation with pentane. Visible absorption maxima (log abs.) 432(5.35), 534(3.99), 575(4.18), 613(3.67) and 676 nm(3.70).

The procedure to produce mono—N—methylated non—*meso*—substituted porphyrins is different in several respects. First, the alkylating agent can be added directly to the solution of the porphyrin in CH_2Cl_2 at the beginning of the reaction. The scale of the reaction is determined by the amount of the porphyrin which will dissolve in refluxing CH_2Cl_2, which varies rather widely. In addition, the chromatographic support medium is different (silica rather than alumina). Typically, the quantity of the porphyrin that can be chromatographed on the same size of column and the eluents that are used for each band are about the same, however. Of course, the colors are quite different, with the mono—N—methylated products typically appearing more brown than burgundy in comparison with the precursor porphyrins. The yields are typically higher, often near 85%. Some of these porphyrins, such as protoporphyrin, are quite light-sensitive and all reactions and separations should be carried out in a darkened fume hood using glassware covered with aluminum foil.

To reduce the amount of time needed for alkylation, of greatest importance for light or oxygen sensitive porphyrins, use of a glass and Teflon^{TM} reaction vessel is recommended. A sturdy glass tube, for example, of the type produced by Fischer and Porter for aerosol sprays, which is rated at 300 psi and is covered with a plastic film, is stoppered with a Telfon^{TM} plug that can be secured by a metal ring.[6] About 500 mg of porphyrin is added to 40 mL of CH_2Cl_2 and a stoichiometric amount of the alkylating agent and a magnetic stir bar added. The vessel is securely sealed and lowered into an oil bath to the level of the solution and heated at 110° C for 1.5 h (use of a shield is recommended). The vessel is then carefully raised from the oil bath and allowed to cool. Before opening, it is best to cool the vessel further with a cold water bath. The reaction mixture is then neutralized and chromatographed as described above. In some cases, as the mixture cools, a precipitate forms and additional CH_2Cl_2 must be added to ensure that all material is dissolved before chromatography.

The yields of mono-N-methylated porphyrins by this technique are higher than by refluxing at normal pressure (about 40% for *meso*-substituted porphyrins and up to 90% for non-*meso*-tetraarylporphyrins such as deuteroporphyrin IX dimethyl ester).

The procedures are the same for ethylfluorosulfonate (ethyl triflate), but the yields are much lower (the best being about 20% for non-*meso*-substituted porphyrins using the pressure vessel). For N-ethylHTPP in CH_2Cl_2, Visible absorption maxima, 432, 531, 573, 613, 675 nm.

Applications. The alkyltrifluoromethanesulfonates are less volatile and apparently less toxic than the alkylfluorosulfonates (although all of these powerful alkylating agents are quite toxic), and are, therefore, the preferred reagents. Currently only the methyl- and ethyltriflates are commercially available, but the method should be applicable to other triflates which can be made from the corresponding alcohol and anhydrous trifluoromethanesulfonic acid (Fluorad, available from 3M Company). Since the larger alkyl chains are less reactive than the methyl derivative, yields on the order of the N-ethylporphyrin or even lower can be expected. This is a fast, simple method for a wide variety of N-methylporphyrins. It is a simple method for synthesizing N-ethylporphyrins, but the yield is much lower.

Methyl Iodide. When H_2TPP and a 100-fold excess of methyl iodide are heated at $100^{\circ}C$ in a sealed tube for 2 days, about 5% is converted to N-CH$_3$HTPP and 15-20% to further methylated products.[3] When anhydrous K_2CO_3 is also added to the reaction mixture, the yield of N, N'-(CH$_3$)$_2$TPP is about 45%.[4] Jackson and coworkers have reported that higher yields can be achieved for reactions of alkyl iodides to produce mono-N-methylated porphyrins by the use of acetic acid in xylene to avoid the formation of multi-alkylated products.[7] Although yields were not reported, it is quite possible that this method is advantageous for synthesis of N-ethyl, N-2-hydroxyethyl and N-propyl derivatives of non-

meso-substituted porphyrins in comparison with the triflate procedure. We have found that the yields of this procedure with *meso*-tetraarylporphyrins, however, are lower than the triflate method.

Procedure.[7] Into a 1.5 L round-bottomed flask containing 1 L of a solution of m-xylene containing 5% glacial acetic acid and 10% methyl iodide is added 500 mg of the porphyrin (*e.g.*, octaethylporphyrin or deuteroporphyrin IX dimethyl ester) and the solution is then refluxed under a nitrogen atmosphere for 30 h. The solution is then cooled, neutralized with 300 mL of 1M aqueous ammonia, and the m-xylene removed using a rotary evaporator. The residue is dissolved in dichloromethane and chromatographed on silica gel.

Applications. N-methyl-, N-ethyl-, N-propyl- and N-2-hydroxyethyl- derivatives of octaethylporphyrin have been reported.[7] Similar derivatives of other non-*meso*-substituted porphyrins should also be feasible by this method. Since alkyl iodides are readily available and not as toxic as some other alkylating agents, this method may also prove quite useful for synthesis of derivatives of non-*meso*-substituted porphyrins using alkyl iodides that are reasonably reactive.

Dimethylsulfate Latos-Grazynski has reported 25-35% yields of N-methyl derivatives of a variety of *meso*-tetra-arylporphyrins (o-Cl, m-Cl, p-Cl, o-CH_3, m-CH_3, p-CH_3 and p-OCH_3) using dimethylsulfate.[8]

Procedure. Into 250 mL of 1,2-dichlorobenzene is added 250 mg of H_2TPP and 1 mL of $(CH_3)_2$. The mixture is refluxed for 1 h, allowed to cool to room temperature, neutralized with solid sodium carbonate and filtered. On a column of acidic alumina, the unreacted porphyrin is removed with 1:10 v/v ethylacetate/toluene and the N-methylporphyrin is removed with chloroform.

Applications. This method has only been applied to H_2TPP derivatives using dimethylsulfate. It is probably applicable to the non-*meso*-substituted porphyrins as well and it may also be feasible with diethylsulfate to produce N-ethylporphyrins. The major advantages of this method are its use of a readily-available and inexpensive reagent and its speed. The yields are better than the methyl iodide method and comparable to the triflate method for H_2TPP derivatives.

Decarboxylation of 2-(N-Porphyrinyl)acetic Acids. The method developed by Callot and coworkers provides relatively high yields of N-methyl- and N-ethylporphyrins and avoids highly toxic reagents.[9] The reaction sequence involves formation of an acetoxy complex of zinc(II) followed by migration of the acetoxy from the zinc atom to a nitrogen atom, acidic elimination of the zinc atom, alkaline hydrolysis of the ester, and photochemical decarboxylation.

Procedure, N-methyltetraphenylporphyrin.[9] A slurry of 500 mg of ZnTPP and 1.5 mL of ethyldiazoacetate in 10 mL of chlorobenzene is heated for 20 h at 100° C. Then 100 mL of CH_2Cl_2 and 3 mL of concentrated HCl are added and the mixture is stirred for 5 min, after which it is neutralized with solid ammonium carbonate, filtered and evaporated. The residue is dissolved in toluene and chromatographed on neutral alumina using toluene to elute unreacted H_2TPP and 5% ethylacetate in toluene to elute N-ethylacetoxyHTPP (65% yield). The ester is refluxed for 3 h with 1 g of NaOH in a solution of 150 mL of THF and ethanol (3:2 v/v). The acid from this step is obtained by diluting the reaction mixture with a solution of 150 mL benzene and 5 mL of acetic acid, and extracting salts and polar solvents with three 200 mL portions of water. After evaporation of the organic solvents using a rotary evaporator, the residue is dissolved in 3 L of CH_2Cl_2 saturated with N_2 and is kept under N_2 while exposed to sunlight for 1 h. After reducing the volume to about 40 mL and adding an equal volume of methanol, the solution is allowed to evaporate to about 30 mL, producing a 50% yield of crystalline product.

Procedure, N-ethyltetraphenylporphyrin.[9] Although this porphyrin can be made using methyl diazopropionate, a higher yield is obtained as follows. A mixture of 500 mg of ZnTPP, 0.46 mL of isopropylamine and 810 mg of methyl pyruvate tosylhydrazone (made in quantitative yield from methyl pyruvate and tosylhydrazine in methanol) are refluxed for 24 h in 30 mL of benzene. After addition of 100 mL of CH_2Cl_2 and 3 mL of concentrated HCl and stirring for 5 min, followed by neutralization with solid sodium carbonate, filtration, evaporation, and chromatography on alumina, the yield of ester is about 35%. Saponification for 3 h at reflux using 100 mL of 3:2 THF and ethanol with 0.75 g of NaOH for 200 mg of ester, followed by evaporation, dissolution in 2 L of CH_2Cl_2 and exposure to sunlight (under N_2) for 1 h gives about 125 mg of N-ethyltetraphenylporphyrin (25% overall yield).

Applications. We have found this method to be highly reproducable and a very useful method for the synthesis of N-ethylporphyrins. It is not limited to H_2TPP derivatives, but the full extent of the reaction has not been investigated. For avoiding the use of very powerful alkylating agents such as "magic methyl" or triflates to make N-methylporphyrins, the use of dimethylsulfate is simpler but requires recycling to match the yield of this method.

σ-Alkylcobalt(III) Porphyrins. Callot and coworkers have optimized conditions for the synthesis of a variety of N-alkyl- and N-arylporphyrins using the migration of a σ-alkyl or σ-aryl group bound to a cobalt(III) atom caused by acidic attack on an oxidized intermediate (Scheme 5.1).[10] The σ-alkyl precursors are made by addition of the corresponding iodide to the Co(I) porphyrin. In the case of *meso*-tetra-arylporphyrins, $NaBH_4$ can be used to reduce the Co(II) porphyrin.[11] Octaethylporphyrin, and other porphyrins whose

Co(II) complexes are more difficult to reduce require 2% sodium amalgam.[12] The reduction can also be accomplished electrochemically. The potential determined in this manner can be a guide for choosing the appropriate chemical reducing agent. The Co(III) σ-alkylporphyrins are very light sensitive and must be handled accordingly.

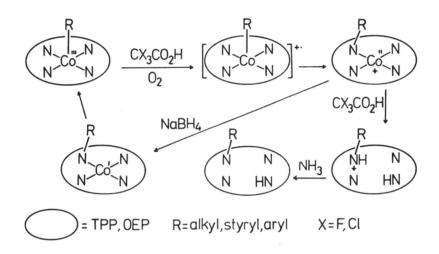

Scheme 5.1

Procedure.[10] To effect the migration, a 100 mg quantity of the σ-alkyl Co(III) porphyrin (thoroughly dried) is dissolved in a solution of 20 mL of CH_2Cl_2 and 2 mL of CF_3COOH. Typically the reaction is complete in 15–30 min (as monitored by alumina TLC) and the solution is then treated with 25 mL of 6M aqueous ammonia, washed with three 100 mL portions of H_2O, dried with magnesium sulfate and chromatographed as described above. The yield of N–CH$_3$HTPP is over 95%, that for N–C_2H_5HTPP is over 85% and that of N–CH$_3$HOEP is over 25%.

Applications. A great advantage of this technique is the fact (demonstrated thoroughly by Callot and coworkers[10], [13] that the migration reaction proceeds intramolecularly. Thus, the N-methyl and N-ethyl derivatives can be produced in good to excellent yield without alkylation of functional groups at the periphery of the porphyrin. (The Co(I) porphyrins are generally extremely strong nucleophiles and the production of the σ-alkyl cobalt(III) complexes is accomplished in a mixture containing pyridine,[11] so peripheral groups potent enough to compete would be very unusual.) In addition, only the mono-N-substituted products are formed. By this method, for example, we have synthesized porphyrins with *meso*-pyridyl groups, such as N-CH$_3$-5-(4-pyridyl)-10,15,20-triphenylporphyrin, visible absorption maxima in CH$_2$Cl$_2$, 434, 533, 576, 612, 674 nm, and N-CH$_3$-5,15-bis(4-pyridyl)-10,20-bis(phenyl)porphyrin visible absorption maxima in CH$_2$Cl$_2$, 433, 529, 571, 613, 671 nm. Since the synthesis of the σ-alkyl cobalt(III) precursors typically proceeds with very good yields, there is a high overall yield with little loss of starting material as multi-alkylated products. This method is excellent for producing N-phenylporphyrins as well as N-alkylporphyrins and can probably be expanded to a wide variety of N-substitutents. The method is limited with respect to the types of σ-bound Co(III) porphyrins that can be made (for example, the σ-benzyl species is not sufficiently stable) and to competition of elimination reactions (*e.g.*, under oxidizing, acidic conditions, σ-CH$_2$CH$_2$Br Co(III) porphyrins undergo elimination to produce ethylene). On the whole, this appears to be the most widely applicable method for N-alkylporphyrin synthesis.

N,N'-Bridged Intermediate. Callot and coworkers have also developed a remarkable method for the synthesis of N-alkyl derivatives of H$_2$TPP which involves the formation of an N,N'-bridged intermediate (described herein for C(H)COOC$_2$H$_5$ as the bridge) formed under phase-transfer conditions. The intermediate is then treated with the alkyl halide to form an N'''-alkyl, N,N'-bridged species that is decomposed with p-toluenesulfonic acid to form the N-alkyl-porphyrin (Scheme 5.2).[14] The bridged intermediate is not indefinitely stable, even in the solid state. We carry out the evaporation of solvent with the rotary evaporator without heating the evaporation flask, employ flash chromatography for purification and use the bridged intermediate as soon as possible. If it is necessary to store the intermediate, it should be refrigerated. Even traces of acid should be avoided until the decomposition step after alkylation of the intermediate.

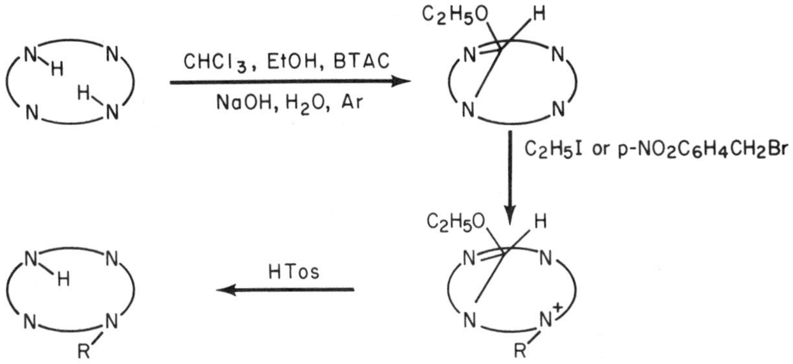

Scheme 5.2

Procedure. A mixture of 500 mg of H_2TPP, 50 mg of triethylbenzylammonium chloride, 500 mg of NaOH, 0.5 mL of H_2O, 0.5 mL of ethanol and 30 mL of $CHCl_3$ (freshly distilled from P_2O_5) is stirred vigorously under N_2 for 2 h. The 5 g of Na_2SO_4 is added and the mixture filtered. The solid is extracted with 3 150 mL portions of CH_2Cl_2 and reduced to about 50 mL using a rotary evaporator. Chromatography on alumina, using CH_2Cl_2 as eluent gives a 25% yields of the N,N'-bridged species (other bridged species can be obtained in comparable yield from a variety of alcohols). (Flash chromatography[15] is highly recommended for this step.) Visible absorption maxima for N,N'-C(H)OC$_2$H$_5$TPP in toluene (log abs.), 434(5.27), 533(4.19), 565 (sh), 608 nm(3.96).

The bridged species (100 mg) is dissolved in CH_2Cl_2 (15 mL) and the alkyl iodide added (1 mL for CH_3I, 4 mL for C_2H_5I) and the mixture stirred at room temperature (overnight for CH_3I, 3 days for C_2H_5I). The solution is then treated with 200 mg of p-toluenesulfonic acid for 4 h at room temperature. After washing with saturated aqueous Na_2CO_3, drying of the organic layer with Na_2SO_4, and chromatography as described above, the overall yield for N-CH_3HTPP and N-C_2H_5HTPP is about 20% (75% – 85% from the bridged intermediate). This reaction can also be used to produce N-p-nitrobenzylHTPP using p-nitrobenzylbromide to decompose the bridged intermediate, giving a 20% overall yield. Visible absorption maxima for N-p-nitrobenzylHTPP in toluene (log abs.), 432(5.46), 495(sh), 527(4.04), 568(4.20), 614(3.65) and 676 nm (3.70).

Applications. This is a fascinating reaction which has advantages over other methods in some situations. Although the formation of the bridge itself does proceed with an excellent yield, attack of this species by alkylating agents is much more facile than attack of ordinary porphyrins. When making N-ethylamine derivatives, for example, we have found no other method described in this chapter to be effective but have obtained good yields using the bridged intermediate and a 2-bromoethylamine precursor (unpublished results).

It appears that acceptable yields are only obtained for the *meso*-tetraarylporphyrins by this method. The only benzyl derivative which we have found to be made readily is p-nitrobenzylHTPP (probably other deactivated benzylbromides could also be used, but the method does not work well with benzylbromide itself.)

Synthesis of N-Phenyl- and N-Vinylporphyrins

N-substituted porphyrins in which the carbon atom bound to the pyrroleninic nitrogen is sp^2 hybridized do not proceed well by direct attack of aryl or vinyl halides on free base porphyrins. For these derivatives, the best preparative methods involve reactions in which a σ-aryl, σ-vinyl or σ-carbene complexes of iron(III) or cobalt(III) are oxidized in the presence of acidμ this causes the migration of the σ-bound group from the metal atom to the nitrogen atom and subsequent displacement of the metal atom. In many cases, nearly quantitative yields have been achieved. This reaction is not only of synthetic interest, but pertains to the reactions of phenylhydrazine with heme proteins *in vivo* as well (as discussed in detail in Chapter 8).

Cobalt(III) σ-Aryl- and σ-Vinylporphyrin Precursors. These reactions involve migration of the σ-bound group from a Co(III) atom to a pyrroleninic nitrogen atom promoted by acid under oxidizing conditions, similar in principle to the migration reactions involving σ-complexes of iron(III) porphyrins (Scheme 5.3). Interesting work concerning these reactions has been published by Ogoshi and coworkers[16] and by Dolphin and coworkers[17], but the most comprehensive studies are those of Callot and coworkers.[10,14,18,19] As discussed above, the reactions are also applicable to the synthesis of N-methyl and N-ethyl porphyrins as well as N-phenyl and N-vinyl porphyrins. The σ-phenyl Co(III) porphyrin precursors are typically made from Grignard reagents and halocobalt(III) porphyrins (in the case of the p-nitrophenyl derivative, Callot and coworkers used p-nitrophenylhydrazine).[10] The synthesis of σ-vinyl cobalt(III) porphyrin precursors was accomplished by the use of the reaction of acetylenes or vinylhalides with Co(I) porphyrins[20] or by the reaction of Grignard reagents or vinyllithium reagents with halocobalt(III) porphyrins.[20] In the case of the σ-styryl precursors, reactions of tosylhydrazones of aryl alkyl ketones with halocobalt(III) porphyrins were used.[19]

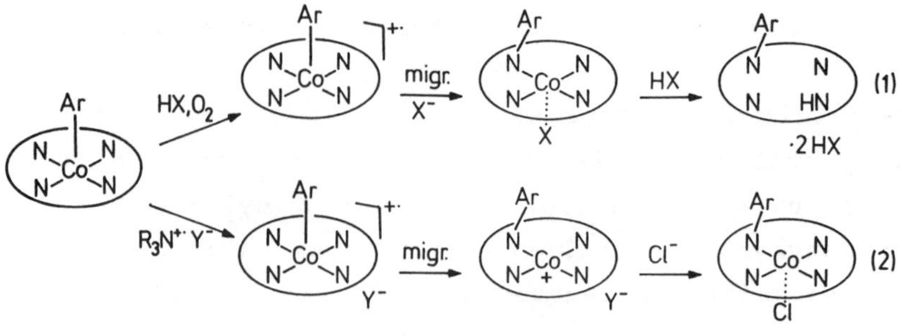

Scheme 5.3

Procedure. The aryl- or vinylcobalt(III) porphyrin is (100 mg) added to a solution of CF_3CO_2H (2 mL) in CH_2Cl_2 (20 mL) at 20°C and the mixture stirred for about 30 min (until alumina TLC indicates completion).[10] The product is then recovered as described previously for the preparation of N-methylporphyrins. N-phenylHTPP and N-phenylHOEP are isolated in about 75% yield and other p-substituted phenyl derivatives were isolated in about 50% yield. N-styrylHTPP was isolated in 16% yield, while $N-C(C_6H_5)=C(H)CH_3$HTPP and $N-C(p-NO_2C_6H_4)=CH_2$HTPP were isolated in 40% and 47% yields, respectively.[10,18]

Applications. This method appears to be the best method for the synthesis of a wide variety of N-phenyl- and N-vinylporphyrins. The method of preparation of σ-aryl and σ-vinyl precursors is quite general and of good yield. The migration reaction itself is rapid and very easy to carry out. The migration is intramolecular, so that it is possible to alkylated the pyroleninic nitrogen without alkylating nucleophilic substituents on the porphyrin periphery. Since the N-phenylporphyrin products are robust, protecting groups that are introduced before the use of phenyllithium or phenylmagnesium bromide can be removed without loss of the N-phenyl substituent.

Iron(III) σ-Aryl Porphyrin Precursors. The oxidative migration of iron(III) complexes has been exploited for the preparation of N-phenyl derivatives of H_2TPP and protoporphyrin IX dimethylester[21,22] as well as the N-vinylidene derivative from the carbene (formed by the reaction of the insecticide DDT with Fe(TPP)Cl).[23,24] In the case of the N-phenyl porphyrins, the precursors are formed by the reactions of phenyllithium[25] or phenylmagnesium bromide[22] with the chloroiron(III) porphyrin precursor.

Procedure, N-phenylHTPP and N-phenylprotoporphyrin IX dimethyl ester. The following is the method of Ortiz de Montellano[21] with the substitution of phenyllithium for phenylmagnesium bromide (Scheme 5.4).[25]. All glassware should be dried under vacuum with a flame and flushed with dry argon immediately before use. In a 100 mL flask containing a magnetic stir bar and fitted with a septum cap, 400 mg of Fe(TPP)Cl and 25 mg of BHT are flushed thoroughly with dry argon. Using a needle lock syringe, 80 mL of dry, deaerated THF (freshly distilled over potassium under argon atmosphere) is added. When 0.25 mL of phenyllithium (2.1 M in pentane) is then added, the solution becomes deep red, indicating the formation of Fe(TPP)Ph. Then a solution containing 120 mL of THF, 200 mL of CH_3OH with 10 mL concentrated H_2SO_4, and 35 mg of BHT is added. The vessel is opened to the air and the mixture is stirred vigorously overnight. It is then washed with 3 100 mL portions of H_2O and extracted with 3 100 mL portions of CH_2Cl_2. The organic layer is washed with aqueous Na_2CO_3 and dried with Mg_2SO_4. The product is best purified by flash chromatography using heptane/CH_2Cl_2 2:1 v/v as eluent. Recrystallization from CH_2Cl_2 gives a yield of about 35%. Visible absorption maxima in CH_2Cl_2, 442, 550, 596, 635, 703 nm.

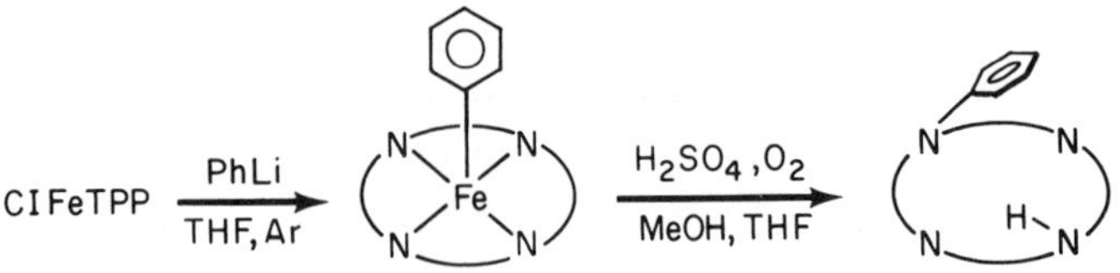

$$CIFeTPP \xrightarrow[\text{THF, Ar}]{\text{PhLi}} \text{(Fe complex)} \xrightarrow[\text{MeOH, THF}]{H_2SO_4, O_2} \text{(N-phenyl porphyrin)}$$

Scheme 5.4

The same method, using phenylmagnesium bromide, has been used to make N-phenylprotoporphyrin IX dimethyl ester in comparable yield.[26] Visible absorption maxima in CH_2Cl_2, 430, 518, 550, 613, 670 nm.

On a smaller scale, N-phenylHTPP has been made in nearly quantitative yield by electrochemical oxidation.[27] The 10 mg of Fe(TPP)Ph was dissolved in a 0.1 M solution of tetra-butylammonium hexafluorophosphate in CH_2Cl_2 (probably about 25 mL) and electrolyzed at 1.0 V vs. S.C.E. (The corresponding Fe(OEP)Ph solution was electrolyzed at 0.8 VC vs. S.C.E.) After electrolysis (indicated by coulometry), the Fe(III) atom was removed with aqueous HCl and the product washed with H_2O. The organic layer was neutralized with aqueous ammonia or collidine, dried with Na_2SO_4 and the solvent removed under reduced pressure. The product was dissolved in diethyl ether or methanol and filtered to remove the electrolyte.

Procedure, N-2,2-bis(phenyl)vinylHTPP and N-2,2-bis(p-chlorophenyl)vinylHTPP. Mansuy and coworkers have described the synthesis of these species from precursors derived from the reaction of either Grignard reagents or DDT with Fe(TPP)Cl.[23, 28]

Over a period of two hours, two equivalents of anhydrous $FeCl_3$ in CH_3CN (5×10^{-4} M) are added anaerobically to 200 mg of the carbene precursor in 200 mL of deaerated toluene at $-20°$ C. The reaction is then allowed to warm to room temperature, a few mL of aqueous 1M HCl is added and the salt and acid are then extracted with water. Aqueous collidine or ammonia is then used to remove traces of acid and the organic layer is washed with H_2O, dried with $MgSO_4$ and reduced in volume using a rotary evaporator. After chromatography on neutral alumina using 9:1 CH_2Cl_2/diethyl ether to remove the N-vinyl band, the product is crystallized from CH_2Cl_2/pentane to give a 60% yield. This procedure gives N-phenylHTPP in 30% yield and N-CH_3HTPP in 5% yield.

Applications. Although the yields for the synthesis of N-phenyl porphyrins are reasonable, higher yields can now be achieved using cobalt(III) precursors. This reaction has been of most interest because of its relation to biological reactions involving phenylhydrazine and DDT. Since the metal-to-nitrogen migration reaction is reversible, the N-substituted porphyrin can be made using Callot's method and iron then inserted to study reaction models for the biological processes.

Synthesis of N-Benzyl- and N-Allylporphyrins.

As described above, N-p-nitrobenzylHTPP can be made by the N,N'-bridged intermediate method of Callot. A much more general method for the synthesis of N-substituted porphyrins in which the N-substituent is a group capable of stabilizing carbocation formation (such as benzyl and allyl groups) has been developed based on the use of diphenylsulfonium salts (Scheme 5.5).[29] These salts can be readily synthesized from diphenylsulfide and the alcohol of the desired substituent.[30] They are typically isolated as crystalline products and are reasonably stable (the more reactive species, such as the benzyl and allyl derivatives, must be stored in a freezer). They are highly reactive alkylating agents from which carbocation-stabilizing groups are readily transferred but the two phenyl substituents are inert.[31]

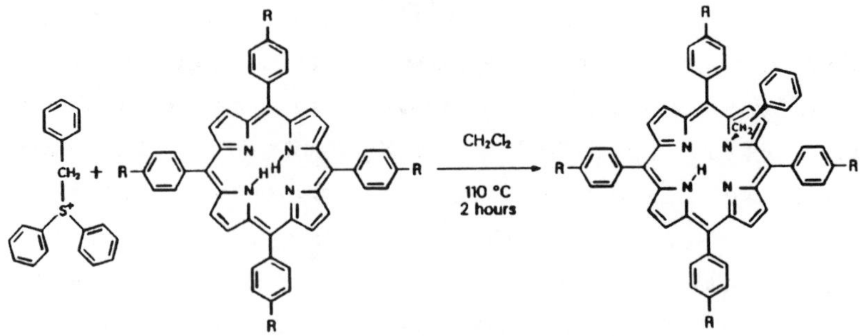

Scheme 5.5

Procedure, N-benzylprotoporphyrin IX Dimethyl Ester.[29]
Since the detailed synthesis of benzyldiphenylsulfonium
tetrafluoroborate is only available in a doctoral thesis[32]
and since all of the substituted benzyl- and allyldiphenylsulfonium salts can be made in an analogous manner, the
procedure is included here.   A typical preparation involves addition of 0.30 mol of methanesulfonic acid to 0.15
mol of deaerated benzyl alcohol at 5°C under argon. After 3
h at ambient temperature, the mixture is washed (three 100
mL portions of 4:1 diethylether/hexane), cooled to 5° C and
converted to the tetrafluoroborate salt by drop-by-drop
addition of 20 mL of H_2O followed by 30 mL of 46% HBF_4.
After extraction with CH_2Cl_2, neutralization with saturated
sodium bicarbonate solution, drying with $MgSO_4$ and reduction
of the volume to about 20 mL, 50 mL of methanol was added,
followed by about 400 mL of diethylether to effect precipitation. Yield 8.6 g (24%) (our results), mp 107-108° C.[32]

For a wide variety of porphyrins, two procedures provide yields in excess of 90%. In the first, protoporphyrin IX dimethyl ester (500 mg) and a 10% excess of benzyldiphenyl-sulfonium tetrafluoroborate (350 mg) in 200 mL of CH_2Cl_2 were stirred at room temperature overnight. In the second, the reagents were combined in the same relative amounts (250 mg of porphyrin and 175 mg of benzyldiphenylsulfonium tetra-fluoroborate in 40 mL of CH_2Cl_2) in a glass tube with Teflon[TM] stopper and heated at $110°$ C for 2h, then cooled to room temperature. In each procedure, the reaction mixture was neutralized with 1 M aqueous ammonia and extracted with water. The product was isolated by column chromatography using silica gel with CH_2Cl_2 and CH_2Cl_2/CH_3CN mixtures as eluents. In both cases, the product before chromatography often appears pure by spectrophotometry and TLC and yields of 95% have been achieved. Visible absorption maxima in CH_2Cl_2 (log abs.), 417(5.04), 511(4.04), 544(3.83), 627(3.34) and 651 nm(3.34).

N-benzyl-5,10,15,20-tetraphenylporphine was synthesized in the same manner, but neutral alumina with CH_2Cl_2 as eluent was used for chromatography. Visible absorption maxima in CH_2Cl_2 (log abs.), 434(5.59), 533(4.11), 573(4.34), 610(3.97) and 672 nm (3.78).

N-allylHTPP was synthesized in the same manner using allyldiphenylsulfonium tetrafluoroborate (made as benzyldi-phenylsulfonium tetrafluoroborate, but isolated as an oil[32]) with a yield of 65% (the visible spectrum is almost identical to that of N-benzylHTPP, unpublished results). This product, unlike any of the others, decomposed slowly (over a period of months) in the solid state. All of the N-benzyl porphyrins (including N-benzyl, N-p-nitrobenzyl and N-p-methylbenzyl derivatives) are stable for many months in the solid state at room temperature.

Synthesis of N,N'-Bridged Porphyrins

One example of this class of compounds has already been discussed: the bridged intermediate used by Callot and co-workers which has the carbon atom of a secondary alcohol bound to a pyroleninic nitrogen atom. It has been used for the synthesis of N-substituted tetraphenylporphyrins.[14] Another porphyrin of this same type, a porphyrin with a saturated carbon atom bridging two neighboring pyrolenininc nitrogen atoms, was reported some time ago by Johnson and coworkers.[33] A second type involves a 1,1'-vinylidene bridge produced from reactions of DDT or similar molecules with iron porphyrins.[23,28,34-37] A third type of bridged species is comprised of 1,2-vinylidene- and 1,2-phenylene-porphyrins (Figure 5.1).[38-40]

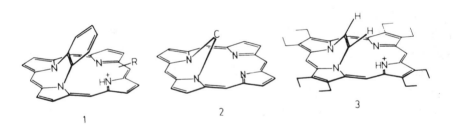

Figure 5.1. Some N,N'-bridged porphyrins. 1) 1,2-phenylene-tetraphenylporphine, 2) with two substituents, the bridged porphyrin of Callot, with a single β-carbon and two β-aryl groups, the DDT derivatives of Mansuy, Balch and others and 3) 1,2-vinylideneoctaethylporphyrin.

N,N'—Bridged Porphyrins With a Saturated Carbon Bridge.
Callot's synthesis involves the direct alkylation of a free
base porphyrin (H$_2$TPP).[14] It is unique in this respect
among preparations of bridged porphyrins, since all of the
others proceed via migration of a moiety bound to a metal
atom (either as a σ—bound species or a complex in which an
organic moiety bridges the metal atom and a nitrogen atom).
The first synthesis of an N,N'—bridged porphyrin was done by
Johnson and coworkers using a scheme in which an ethoxycarbonylmethylene group bound simultaneously to a cobalt atom
and a pyrrolenic nitrogen forms a bond to the neighboring
nitrogen atom.[33]

*Procedure, N,N'-ethoxycarbonylmethyleneoctaethylporphy-
rin.* To form the bridged intermediate, NCo-ethoxycarbonyl-
methylene cobalt(III) octaethylporphyrin chloride, 0.5 mL of
ethyl diazoacetate is added to 500 mg of cobalt(II) octaethylporphyrin in 40 mL of chloroform and stirred for 20 min
at room temperature, giving a color change from red to
brown. The solvent is removed with a rotary evaporator, the
residue dissolved in 20 mL of CHCl$_3$ and 10 g of polypro-
lactam added. The solvent is then removed again and the
residue is dissolved in petroleum ether and chromatographed
on polylactam (4 x 100 cm column). The unreacted H$_2$OEP is
removed first with petroleum ether containing 0.5% acetone
and the bridged complex removed second with 5% acetone in
petroleum ether (382 mg, 64% after crystallization from
CH$_2$Cl$_2$/petroleum ether). This complex is then dissolved in
a minimum amount of CH$_2$Cl$_2$ and added to a solution of 10%
HCl in dry ethanol (10 mL/ 100 mg of complex) and stirred
for two hours, while the color changes from brown to green.
The reaction mixture is then poured into saturated sodium
hydrogen carbonate solution (an excess) and extracted into
CH$_2$Cl$_2$. The extract is washed with NaHCO$_3$ solution, then
with water, dried, and evaporated. The product, N,N'—etho-
xycarbonylmethyleneoctaethylporphyrin, was isolated as the
major product using preparative silica TLC (benzene/acetone
3:2 v/v), The yield was not reported.[41] Visible absorption

maxima (log abs.) 239(4.66), 297(4.05), 395(5.20), 525(3.88), 555(4.02), and 598 nm(3.79).

Applications. This does not appear to be a general method for formation of N,N'-bridged porphyrins. While the diazoacetates react with cobalt(III) porphyrins to give a metal–nitrogen bridged species, other diazo compounds react with the cobalt(II) porphyrins to give alkylcobalt(III) porphyrins[42] which, under oxidizing conditions, undergo acid-promoted migration to give mono–N–alkyl porphyrins or non–N–alkylporphyrins.[10] One positive aspect to this synthesis is that it might be possible to modify this bridging group to make other bridged derivatives. Reaction conditions would have to be delicate, however. Like the bridge resulting from Callot's method, the ethoxycarbonylmethylene bridge is not very stable: treatment with base removes this bridging group and treatment with acid produces the mono–N–ethoxycarbonylmethyleneporphyrin.[33]

Synthesis of N,N'–1,1'–Vinylideneporphyrins. Compounds of this type have all been synthesized from iron(III) porphyrin precursors. The precursors have included chloroiron(III) complexes of tetraphenyl– and tetrakis(p-anisyl)porphyrin and the carbene sources 1,1,1–trichloro–2,2–bis(p-chlorophenyl)ethane (DDT), 1,1,1–trichloro–2,2–bis(p-methoxyphenyl)ethane (DMDT) and 1,1,1–trichloro–2,2–diphenyl–ethane.[28,37] The first product of these reactions is a complex with a carbene moiety σ–bound to the iron atom, produced by the reductive dechlorination of a trichloroethane molecule in the presence of an iron(II) porphyrin. Migration proceeds via two steps first forming a complex with the carbene interposed between the iron atom and a pyrroleninic nitrogen (by one-electron oxidation of the carbenic complex) and then forming an N,N'–carbene–substituted product by further oxidation.

Procedure, $N,N'-C=C(p-ClC_6H_4)_2$ tetraphenylporphyrin.[23]
The Fe(II) carbene precursor complex is made by adding 1.4
equivalents of DDT over a period of three hours to a deaerated, argon saturated, stirred solution of CH_2Cl_2/CH_3OH,
9:1 v/v, 10^{-3} M in TPP-Fe(II) containing an excess of iron
powder. The solution is then filtered, washed with a dilute
aqueous solution of sodium dithionite, and crystallized from
CH_2Cl_2/CH_3OH (50% yield after three recrystallizations[34]).
The iron-nitrogen bridged intermediate is then formed as
follows: to a solution of the carbene complex in benzene
(10^{-4} M), 1 equivalent of anhydrous $FeCl_3$ in CH_3CN (0.10 M)
is added over a period of five hours. The solvent is then
removed with a rotary evaporator and the product isolated by
column chromatography using silica gel with benzene to remove the unreacted carbene complex and benzene/acetone,
70:30 v/v, to remove the bridged complex (80% yield[43]).
Reaction of this complex (0.01 M in CH_2Cl_2) with 1.1 equivalents of anhydrous $FeCl_3$ at $20^\circ C$ for 30 min after washing
with water, drying of the organic layer and column chromatography using silica gel with ethyl acetate/CH_2Cl_2 3:1 v/v
as eluent gives $N,N'-C=C(p-ClC_6H_4)_2$ HTPPCl in 80% yield.[23]
Visible absorption maxima in benzene (log abs.), 430(5.15),
509(3.89), 549(4.08), 585(4.19) and 631 nm(3.85).

Gold and coworkers prepared similar compounds without
isolating the bridged intermediate. For the preparation of
$N,N'-C=C(p-ClC_6H_4)_2$ tetrakis(p-anisyl)porphyrin and 2 mL of
9:1 degassed CH_2Cl_2/CH_3OH containing 6 mg of DDT was added
over 3 h to 5 mL of 9:1 CH_2Cl_2/CH_3OH under N_2 containing 10
mg of tetraanisylporphinatoiron(III) chloride and an excess
of iron powder (150 mg). The reaction mixture was filtered
and the solution evaporated under N_2 and redissolved in 1 mL
CH_2Cl_2 to which was added 10 mg of anhydrous $FeCl_3$. After a
few minutes, the color change from brown to green had ceased
and the solution was filtered, washed and chromatographed
first by column chromatography using neutral alumina ($CHCl_3$
to remove unreacted material followed by 1:4 $CH_3OH/CHCl_3$ to
remove the product) and, finally by preparative TLC using
neutral alumina (with 1:4 $CH_3OH/CHCl_3$ as eluant), giving an

overall yield of 52%.[37] Visible absorption maxima (log abs.)
453(5.02), 560(3.82), 605(4.29) and 645 nm(3.90).

Applications. The extent of this type of migration reaction has not been explored fully. Mansuy and coworkers also
produced N,N'-vinylidene products by electrochemcial oxidation of the iron-nitrogen bridged intermediate, but not on a
preparative scale and only for the type of species discussed
above. The initial complex in the reaction sequence, consisting of the dechlorinated organic precursor σ-bound to an
iron(II) porphyrin moiety, has been made with a variety of
para-substituted *meso*-tetraphenylporphyrins, octaethylporphyrin and protoporphyrin IX dimethyl ester,[34] so it
seems likely that this reaction could be extended to a
variety of porphyrin ring systems. The 2,2'-bis(phenyl)-
N,N'-1,1'-vinylidene bridge may be especially stable and it
is not yet clear if other, less closely related porphyrins
with a single atom of a vinylidene group bridging neighboring nitrogen atoms will be reasonably stable.

Synthesis of N,N'-1,2-Vinylidene- and
N,N'-1,2-Phenyleneporphyrins.

Although nature uses iron porphyrins for the synthesis of
the two-atom N,N' porphyrin bridge (from the reaction of 1-
aminobenzotriazole with cytochrome P-450[43,44]), the synthetic procedures reported by Dolphin and coworkers,[38] Callot
and coworkers[39,40] and Setsune and coworkers[45] used
cobalt(III) porphyrins. Dolphin's synthesis of N,N'-(1,2-
vinylidene)OEP uses a cobalt(III) porphyrin simply as a
means of producing the precursor, N-formylmethylHOEP, which
is then converted to the 1,2-vinylidene derivative by treatment with acid. Callot's method, however, involves the
formation of an N-vinylporphyrin complex of cobalt(II)
followed by oxidation, electron transfer to the N-vinyl
moiety, and then intramolecular attack of the 2-vinylidene

carbon on the vicinal nitrogen atom to form the 1,2-vinyli-
dene bridge. Setsune has produced a variety of 1,2-vinyli-
dene bridged species including N,N'-(1,2-vinylidene)HOEP$^+$,
as well as species with phenyl, alkyl and hydroxymethyl
substituents, by addition of the corresponding alkyne to a
solution of Co(OEP)(H$_2$O)$_2$ClO$_4$ in the presence of an
oxidizing agent.

Procedure, N,N'-(1,2-vinylidene)octaethylporphyrins.[46]
The initial complex, Co(OEP)(H$_2$O)$_2$ClO$_4$, is made by the
method of Sugimoto, *et al.* (*Bull. Chem. Soc. Jpn.,* **54** (1981)
3425). It is dissolved in CH$_2$Cl$_2$ with one to two equivalents of FeCl$_3$ and stirred for several minutes under an
atmosphere of acetylene until the color change to reddish-
green has ceased (the progress of the reaction is readily
monitored by TLC). After the reaction mixture is washed
with 10% aqueous HClO$_4$, dried and chromatographed on silica
gel using 5:1 CHCl$_3$/acetone, the yield is 52%. Visible
absorption maxima of the monocation (log abs.), 393(5.07),
535(3.89), 570(4.00) and 615 nm(3.19). The yield for the
corresponding N,N'-(1,2-bis(phenyl)vinylidene)HOEP$^+$ was 71%.
Visible absorption maxima for monocation (log abs.),
401(5.13), 535(3.94). 569(4.06), and 614 nm(3.63). Yields
for species with one n-butyl, one –CH$_2$OH, one phenyl or two
–CH$_2$OH groups were all about 50%.

Applications. This method appears to be general for a
variety of substituents. It is necessary to use the diaquo-
cobalt(III) precursor rather than, for example, the bromoco-
balt(III)octaethylporphyrin. According to the authors, this
may be due to quite different reactivities of the different
type of cation radical formed upon oxidation. Thus, exten-
sion to porphyrins other than octaethylporphyrin may depend
on the nature of the cation radicals that they form. This
is an extremely simple and efficient method deserving
further exploration. It is important to note that the
bridged products are not very stable in the presence of base
and are best isolated as the monocationic salts. Since this

method has not yet been extended to the *meso*-tetraaryl-
porphyrin series, an alternate synthesis is given below.

Procedure, N,N'-(1,2-bis(phenyl)vinylideneHTPP (Scheme
5.6).[46] To 140 mg of Co(II)-σ-1,2-bis(phenyl)vinyli-
deneHTPP[19] in 50 mL of CH_2Cl_2 is added 300 mg of (p-Br-
$C_6H_4)_3N^+SbCl_6^-$ and the solution stirred for 10 min at room
temperature. It is then concentrated using a rotary
evaporator and chromatographed on a column of silica gel (70
g) using 5% methanol in CH_2Cl_2 as eluent. After removal of
the solvent and crystallization from CH_2Cl_2/hexane the yield
is about 20%. Visible absorption maxima in CH_2Cl_2, (log
abs.) 430(5.23), 554(4.00), 592(4.18) and 640 nm(3.79). For
the monoprotonated salt, 426(5.16) 556(3.95), 588(4.12) and
638 nm(3.80).

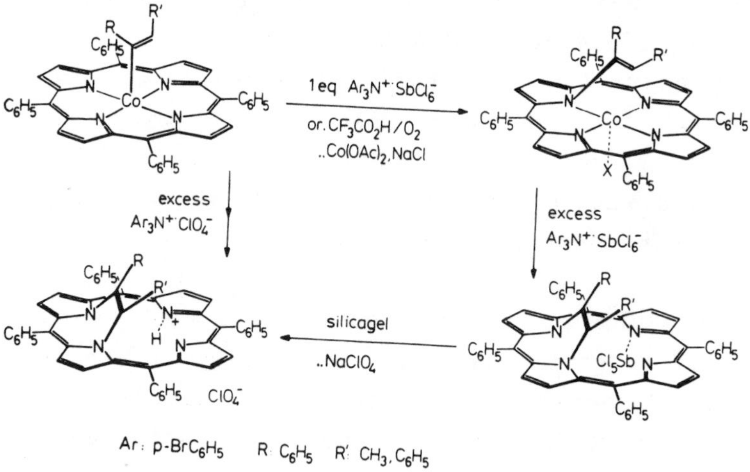

Scheme 5.6

Procedure, N,N'-(1,2-phenylene)HTPP.[40,46] A solution of 643 mg of Co(N-PhTPP)ClO$_4$ and 1.5 g of (p-Br-C$_6$H$_4$)$_3$N$^+$ClO$_4^-$ in 55 mL of 1,1',2,2'-tetrachloroethane was refluxed under argon for 70 h. The reaction mixture is then allowed to cool to about 60^{0}C and at least a 10-fold molar excess of Co(OAc)$_2$ dissolved in a minimum of hot methanol is added. Metallation is immediately evident by a color change to deep green. The solution is cooled to room temperature, adsorbed on a column of silica gel (50-60 g /100 mg of starting material) and the starting material eluted with 2% methanol on CH$_2$Cl$_2$. A 5% solution of methanol then removes the product. Solvent is removed using a rotary evaporator, the residue dissolved in methanol and precipitated with a saturated aqueous solution of NaClO$_4$. After washing with water and air-drying, the product is crystallized from CH$_2$Cl$_2$/hexane, giving 394 mg, a 66% yield with 95 mg of the starting material recovered. Visible absorption maxima (log abs.) 442(4.97), 566(3.89), 602(4.08) and 656 nm(3.80). Derivatives with 4-OCH$_3$ or 4,5-(OCH$_3$)$_2$ substituents were prepared similarly with yields of 33% and 38%, respectively. The N,N'-(1,2-phenylene)octaethylporphyrin analog was not obtained this procedure, but was made in a 3% yield using electrochemical oxidation of Co(N-phenylOEP)ClO$_4$.[40,46] Visible absorption maxima of the monoprotonated salt (log abs.) 410(4.89), 542(3.83), 576(3.94) and 624 nm(3.40).

Applications. This method has some generality, but several important limitations as well. The synthesis of N,N'-styryl derivatives, where the carbon atom bound to the pyroleninic nitrogen bears a phenyl group, gives similar yields to those of the TPP derivatives (about 20%), and, therefore, it may be general for non-*meso*-substituted as well as *meso*-substituted porphyrins. But the reaction only proceeds for such phenyl-bearing derivatives, not when an unsubstituted vinyl or α-styryl group is attached to the nitrogen atom in the starting material. (Callot has proposed that positive charge stabilization by the N-substituent is essential to the mechanism, favoring $-CH_3$ and phenyl groups.)

At present, this is the only method described for the formation of phenylene derivatives, but it only appears useful for *meso*-tetraarylporphyrins. It is interesting that Callot and Cromer also synthesized N,N'-(1,2-napthylene)TPP in good yield by this method.[39]

Synthesis of Metal-Nitrogen Bridged Porphyrins

The two types of metal-nitrogen bridged porphyrins for which several examples exist are those in which the bridging moiety consists of an alcoxycarbonylmethylene moiety and those with a tosylamino bridge (see Figure 2.8). The synthesis of the former type dates from the work of Johnson and his coworkers in 1975[33] and includes cobalt complexes of octaethyl- and *meso*-tetraphenylporphyrin with an ethoxycarbonylmethylene bridge,[33,47] and of tetraphenylporphyrin with an ethoxycarbonylmethylmethylene group (made directly from the Co(II) complexes by addition of diazoacetate or diazopropionate, respectively),[48] and a nickel tetraphenylporphyrin complex (this was made by inserting nickel into N-ethoxycarbonylmethyleneHTPP followed by reaction with base, which removes one α-proton to form the bridged product).[49] The complexes with an N-tosylamino bridge include those of nickel and mercury, made by inserting the metal atom into N-tosylaminoHTPP[49,50] and an iron complex made from the reaction of a tosylnitrene with Fe(TPP)Cl.[51] In all cases, the metal atoms can be removed to give the N-substituted free base porphyrins. In the case of cobalt, the bridged complexes can also be converted into σ-alkylcobalt(III) porphyrins. These bridged species represent relatively stable intermediates between the N-alkylporphyrin complexes and σ-alkylmetalloporphyrins. The isolation of bridged species, however, appears to be limited to only a few metals.

Procedure, Co(III), Co(II), Ni(II) and Hg(II) N-metal bridged complexes of N-ethoxycarbonylmethyleneoctaethylporphyrin. The preparation of the Co,N-ethoxycarbonylmethyleneoctaethylporphinatocobalt(III) complex has been described above as the intermediate in the synthesis of the free base. To obtain the bridged Co(II) complex, the Co(III) complex is reduced by using a syringe to add 1.5 mL of 0.15 M $CrCl_2$ in ethanol syringe to a deaerated solution in CH_2Cl_2 (5 mL/ 100 mg of complex) under N_2 or Ar and stirring for 5 min (the color changes from brown to red–brown). The solution is then poured into aqueous saturated sodium hydrogen carbonate (100 mL/100 mg of complex) and extracted with three 100 mL portions of CH_2Cl_2. The extracts are washed with water, dried with $MgSO_4$, reduced in volume with a rotary evaporator and chromatographed on silica gel using 3% acetone in petroleum ether to remove nonpolar impurities and 20% acetone in petroleum ether to elute the product. (66% yield after crystallization from acetone with 1% aqueous sodium chloride[33]) Visible absorption maxima (log abs.) 382 (4.68), 426(4.98), 542(3.89) and 585 nm(4.06).

The chloro-mercury(II) complex is made by mixing the free base, made by acid displacement of the Co(II) atom as described previously, (about 50 mg in 10 mL of CH_2Cl_2) to a solution of $HgCl_2$ in methanol (about 130 mg/ 10 mL followed by the addition of solid sodium acetate (100 mg/50 mg of the free base). Crystals were grown from $CHCl_3/CH_2Cl_2/CH_3OH$.[53]

Procedure, Chloro-(Co,N-ethoxycarbonylmethylenetetraphenylporphyrin.[54] Ethyl diazoacetate (0.24 mL) is added to CoTPP (300 mg) in 50 mL of chloroform and stirred for 20 min and evaporated quickly using a rotary evaporator. Then 50 mL of pentane is added and the mixture evaporated rapidly. The precipitate is washed with pentane, dissolved in 20 mL of CH_2Cl_2 and reprecipitated and dissolved in like manner two more times to give a 60% yield.

The Co(II) complex and the free base can be produced as described above for the octaethylporphyrin species, giving

45% and 85% yields, respectively. Note that the free base, N-ethoxycarbonylmethylHTPP, can be produced more readily by the reaction of ZnTPP with ethyl diazoacetate (66% yield)[9], as described above for Callot's synthesis of N-CH$_3$HTPP.

Procedure, Ni,N-ethoxycarbonylmethylenetetraphenylporphyrin.[50] The non-bridged Ni(N-ethoxycarbonylmethylene)TPP complex is made by adding a solution of nickel acetate in warm methanol (500 mg to 50 mL) to a refluxing solution of the 150 mg of the free base in 40 mL of CHCl$_3$. After about 30 min, the CHCl$_3$ is boiled off and the complex precipitates from the methanolic solution. The bridged complex is formed with an overall yield of about 50% by stirring a solution of 100 mg of the non-bridged complex with 0.5 g of Na$_2$SO$_4$ and 5 mL of triethylamine in 5 mL of CH$_2$Cl$_2$ for 3 days at 25°C. The solution is then filtered, the solvent evaporated and the residue chromatographed on silica gel. The product is eluted with toluene and crystallized from CHCl$_3$/CH$_3$OH.

Procedure, Ni,N-tosylaminotetraphenylporphyrin.[50] The free base was obtained by leaving a solution of 0.5 g of ZnTPP and 2.0 g of p-toluenesulfonyl azide in six stoppered 100 mL erlenmeyer flasks in the sunlight for 8 h, then combining the solutions and adding 5 mL of 16N HCl with vigorous shaking, following by treatment with excess solid ammonium carbonate for 30 min. The slurry was filtered and the solvent evaporated to dryness using a rotary evaporator. The mixture was dissolved in CH$_2$Cl$_2$ and chromatographed on an alumina column (250 g), giving first H$_2$TPP and then the product using CH$_2$Cl$_2$ as eluent. (50% yield, Visible absorption maxima (log abs.) 432(5.34), 510(3.36), 548(3.97), 585(4.00), and 640 nm(3.90). The complex was obtained as described above for the non-bridged N-ethoxycarbonylmethylene complex, but in this case, the bridged species is formed. Visible absorption maxima (log abs.) 425(5.10), 566(4.07) and 585 (sh, 4.03).

Procedure, Chloro-Fe,N-tosylaminotetraphenylporphyrin.[51]
A solution of 70 mg of Fe(TPP)Cl in 10 mL of CH_2Cl_2 is
deaerated with argon and added via syringe to an argon-
flushed flask containing 150 mg of PhI=NTs (prepared by the
method of Yamada, *et al., Chem. Letters* (1975) 361-362).
After 15 min, the solution is filtered and the complex
precipitated with pentane and recrystallized from 2:1
CH_2Cl_2/CH_3OH, with an 80% yield. Visible absorption maxima
422, 520, 560 and 690 nm.

Applications. These appear to be highly specific reactions which may be limited a few metal ions since these
species appear to represent intermediates between two types
of complexes that are often more stable than the bridged
species itself. One of the most useful aspects of these
species is, in fact, their role as mechanistic models.

Synthesis of N-Aminooctaethylporphyrin and
N-Aminotetraphenylchlorin

In an approach like that used for the synthesis of N-
alkylporphyrins with sulfonyl esters, Callot has synthesized
N-amino derivatives using O-mesitylsulfonylhydroxylamine,
RSO_2ONH_2.[54] He has also shown that the N-amino species can
be converted readily into the corresponding N-tosylamino
compounds by reaction with tosyl chloride in the presence of
base. It is interesting that the product in the case of the
reaction with tetraphenylporphyrin is not N-aminoHTPP, but
N-aminotetraphenylchlorin, possibly arising because the
chlorin is more reactive toward amination and the chlorin
can be produced from the reaction of H_2TPP with O-mesitylsulfonylhydroxylamine.[54] The N-amino compounds are not as
stable as the N-tosylamino compounds, decomposing to the
non-N-substituted porphyrin or chlorin when treated with
Cu(II), Ni(II) or even Zn(II). By contrast, octaethylporphyrin N-oxide does not survive alumina chromatography[55],
but does form stable complexes of Ni(II) and Cu(II).[56]

Procedure, N-aminooctaethylporphyrin. Equal amounts of O-mesitylsulfonylhydroxylamine (prepared by the method of Yamada, *et.al., Synthesis,* (1977) 1) and octaethylporphyrin (0.5 g) are stirred for 20 h in $CHCl_3$ at room temperature. The solvent is evaporated and the solid residue dissolved in CH_2Cl_2 and chromatographed on alumina (a column of 100 g). The first band is H_2OEP (43%) and the second band, which is green, is $N-NH_2HOEP$ (24% yield, or 40% based on recovered H_2OEP).[54] Visible absorption maxima (log abs.) in toluene, 408(5.15), 510(3.98), 545(3.76), 573(3.83) and 628 nm(3.77).

Procedure, N-amino-5,10,15,20-tetraphenylchlorin. Equal amounts of O-mesitylsulfonylhydroxylamine and H_2TPP (275 mg) are stirred in $CHCl_3$ at room temperature for 24 h, the solvent evaporated and the residue dissolved in CH_2Cl_2 and chromatgraphed on alumina. About 70% of the H_2TPP is recovered unchanged (the first band from the column) and N-aminotetraphenylchlorin is obtained in 10% yield.[54] The reaction proceeds 2-3 times faster if tetraphenylchlorin is used as starting material, but the most convenient starting material is crude H_2TPP (not purified with DDQ). Visible absorption maxima (log abs.) in toluene, 428(5.09), 447(sh), 545(4.09), 610(sh), and 660 nm(3.92).

Applications. This is, of course, a highly specific procedure, but it does show the general utility of sulfonyl precursors which are excellent leaving groups for electrophilic addition reactions. The procedure appears to apply to a variety of porphyrin precursors, but since it leads to the N-aminochlorin product in the case of the tetraphenylporphyrin reaction, it can be expected to lead to chlorins of other relatively easily reduced porphyrin precursors. The N-aminoporphyrins or chlorins are readily converted to the more stable N-tosylamino derivatives. The replacement of the amino proton may perhaps be exploited for the synthesis of other N-amino derivatives. The conversion of an N-aminochlorin to an N-aminoporphyrin may be difficult because N-aminoporphyrins and chlorins are relatively unstable.

Total Synthesis of Isomers of N-Methylporphyrins

N-Methylprotoporphyrin IX (as well as some other N-alkyl-porphyrins) has been shown to be a powerful inhibitor of the enzyme ferrochelatase. This porphyrin and several related porphyrins are formed from the destruction of cytochrome P-450 by a number of drugs, xenobiotics and natural compounds even as simple as ethylene (see chapters on ferrochelatase inhibition and cytochrome P-450 reactions, Chapters 6 and 7, respectively, for a full discussion). The synthetic alkylation of protoporphyrin IX and other nonsymmetrically substituted porphyrins leads to a random mixture of the possible isomers, 25% alkylation of each of the four pyrrolenine rings. Although it is possible to separate these isomers by HPLC, the procedure is quite tedious and the amounts of material that can be isolated are limited. This may not be a serious problem if only a small amount of material is required, but it would be useful for some studies to have access to a reasonably large amount of each isomer in pure form.

In 1985, Jackson and coworkers published the first total synthesis of an N-alkylporphyrin. Their procedure involved the condensation of an N-methylated pyrromethane and a diformylpyrromethane, followed by oxidation by aeration and reesterification (the bis(pyrrolenine) reagents and the reaction product are shown in Figure 5.2)[57] The overall yield of the N-methyl-1,2,3-triethyl-4,5,8-trimethyl-6,7-methylpropionateporphine was a remarkable 25%. Although limited in scope, the relatively high yield of this synthesis and the fact that it was the first successful total synthesis of an N-alkylporphyrin related to natural porphyrins are noteworthy accomplishments.

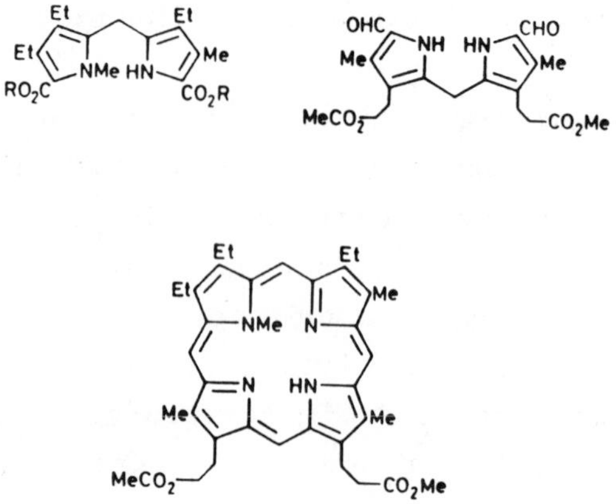

Figure 5.2. The N-methylated pyrromethane and diformylpyrromethane (upper left and right, respectively) used by Jackson and coworkers for the total synthesis of the N-methyl-1,2,3-triethyl-4,5,8-trimethyl-6,7-methylpropionateporphine (lower) (Ref. 57).

The procedure for obtaining each of the four isomers (as racemic mixtures) of N-methylprotoporphyrin IX published by Smith and Pandey is of direct interest with regard to biochemical studies of N-alkylporphyrins.[58] Since this is an involved synthetic scheme that is well explained in the original article, it is not given in detail here.

Synthesis of Metal Complexes

There has been relatively little work in this area in contrast with the extensive range of synthetic methods that have been developed for the non-N-substituted porphyrins (see, for example, ref. 59). Because the N-substituted porphyrins often react much more rapidly with metal complexes than the non-N-substituted complexes, it is possible to synthesize derivatives of a number of transition metal ions by simply mixing solutions of organic solvents such as dichloromethane, tetrahydrofuran, methanol, acetone or acetonitrile containing the N-substituted porphyrin and the another containing a chloro-, bromo-, nitrato-, acetato- or other relatively labile and soluble complex of the metal ion and stirring for a few minutes at room temperature. This simple method is appropriate for complexes of Zn(II), Co(II), Mn(II), Cd(II), Hg(II), Ni(II) and Cu(II). For complexes of Pd(II), which often readily undergo nucleophilic displacement of the N-substituent, it is necessary to use lower temperatures and avoid nucleophilic solvents. For the complexes of Cd(II) and Hg(II) it is necessary to scrupulously avoid acid. Complexes of Fe(II) are best made by refluxing a solution containing anhydrous $FeCl_3$ or $FeCl_2$ and an excess of iron wire (readily removed with the magnetic stir bar) in refluxing THF with a small amount of non-coordinating base.[60] For these procedures, little excess of the metal ion is typically necessary and any excess can be removed by evaporating the solution and dissolving the porphyrin in dry dichloromethane or, for those complexes which are not sensitive to acid, by an extraction using dichloromethane and water.

The methods which have been used for the formation of metal complexes have all been straightforward and it is likely that more imaginative methods will soon be developed.

References

1. W.K. McEwen, J. Amer. Chem. Soc., **58** (1936) 1124-1129.

2. G.R. Dearden and A.H. Jackson, Chem. Comm., (1970) 205-206.

3. D.K. Lavallee and A.E. Gebala, Inorg. Chem., **19** (1974) 2004-2008.

4. H.M.G. Al-Hazimi, A.H. Jackson, A.W. Johnson and M. Winter, J. Chem. Soc., Perkin I, (1977) 98-103.

5. A. Neuberger and J.J. Scott, Proc. Roy. Soc., **A213** (1952) 307-326.

6. D.K. Lavallee and M.J. Bain, J. Chem. Educ., **53** (1976) 221.

7. J.A.S. Cavaleiro, M.F.P.N. Condesso, A.H. Jackson, M.G.P.M.S. Neves, K.R.N. Rao and B.K. Sadashiva, Tetrahedron Lett., **25** (1984) 6047-6048.

8. L. Latos-Grazynski, Inorg. Chem., **24** (1985) 1681-1686.

9. H.J. Callot, Tetrahedron Lett., **33** (1979) 3093-3096.

10. H.J. Callot, F. Metz and R. Cromer, Nouv. J. Chim., **8** (1984) 759-763.

11. M. Perrée-Fauvet, A. Gaudemer, P. Boucly and J. Devynck, J. Organomet. Chem., **120** (1976) 439-451.

12. H. Ogoshi, E.I. Watanabe, N. Koketsu and Z.I. Yoshida, Bull. Chem. Soc. Jpn., **49** (1976) 2529-2436.

13. H.J. Callot and R. Cromer, Nouv. J. Chim., **8** (1984) 765-770.

14. H.J. Callot, J. Fisher and R. Weiss, J. Am. Chem. Soc., **104** (1982) 1272-1276.

15. W.C. Still, M. Kahn and A. Mitra, J. Org. Chem., **43** (1978) 2923-2925.

16. H. Ogoshi, E.I. Watanabe, N. Koketzu and Z.I. Yoshida, J. Chem. Soc., Chem. Commun., (1974) 943-944.

17. D. Dolphin, D.J. Halko and E. Johnson, Inorg. Chem., **20** (1981) 4348-4351.

18. H.J. Callot and E. Schaeffer, Tetrahedron Lett., **21** (1980) 1335-1338.

19. H.J. Callot and E. Schaeffer, J. Organomet. Chem., **193** (1980) 111-115.

20. H.J. Callot and E. Schaeffer, Tetrahedron Lett., **3** (1977) 239-242.

21. P.R. Ortiz de Montellano, K.L. Kunze and O. Augusto, J. Am. Chem. Soc., **104** (1982) 3545-3546.

22. P.R. Ortiz de Montellano, H.S. Beilan, K.L. Kunze and B.A. Mico, J. Biol. Chem., **256** (1981) 4395-4399.

23. M. Lange, and D. Mansuy, Tetrahedron Lett., **22** (1981) 2561-2564.

24. J.-P. Battioni, J.-C. Chottard and D. Mansuy, Inorg. Chem., **21** (1982) 2056-2062.

25. D.K. Lavallee and D. Kuila, Inorg. Chem., **23** (1984) 3987-3992.

26. P.R. Ortiz de Montellano and K.L. Kunze, J. Am. Chem. Soc., **103** (1981) 6534-6536.

27. D. Lancon, P. Cocolios, R. Guilard and K.M. Kadish, J. Amer. Chem. Soc., **106** (1984) 4472-4478.

28. D. Mansuy, J.-P. Battioni, D. Dupré, E. Sartori and G. Chottard, J. Am. Chem. Soc., **104** (1982) 6159-6161.

29. D.K. Lavallee, A. White, A. Diaz, J.-P. Battioni and D. Mansuy, Tetrahedron Lett., **27** (1986) 3521-3524.

30. B. Badet and M. Julia, Tetrahedron Lett., **13** (1979) 1101-1104.

31. B. Badet, M. Julia and C. Lefebvre, Bull. Soc. Chim. Fr., (1984) II 431-434.

32. B. Badet, These, Univ. Pierre et Marie Curie, 1980.

33. A.W. Johnson, D. Ward, P. Battan, A.L. Hamilton, G. Shelton and C.M. Elson, J. Chem. Soc., Perkin I, (1975) 2076-2085.

34. D. Mansuy, M. Lange and J.-C. Chottard, J. Am. Chem. Soc., **100** (1978) 3213-3214.

35. D. Mansuy, M. Lange and J.-C. Chottard, J. Am. Chem. Soc., **101** (1979) 6437-6439.

36. B. Chevrier, R. Weiss, M. Lange, J.-C. Chottard and D. Mansuy, J. Am. Chem. Soc., **103** (1981) 2899-2901.

37. T.J. Wisnieff, A. Gold and S.A. Evans, Jr, J. Am. Chem. Soc., **103** (1981) 5616–5618.

38. J. Setsune and D. Dolphin, Organomet., **3** (1984) 440–443.

39. H.J. Callot and R. Cromer, Tetrahedron Lett., **26** (1985) 3357–3360.

40. H.J. Callot, R. Cromer, A. Louati and M. Gross, J. Chem. Soc., Chem. Commun., (1986) 767–769.

41. H.J. Callot and E. Schaeffer, J. Organomet. Chem., **145** (1978) 91–99.

42. D. Mansuy, J.P. Battioni, J.-C. Chottard and V. Ullrich, J. Am. Chem. Soc., **101** (1979) 3971–3973.

43. P.R. Ortiz de Montellano and J.M. Mathews, Biochem. J., **195** (1981) 761–764.

44. P.R. Ortiz de Montellano, J.M. Mathews and K.G. Lanory, Tetrahedron, **40** (1984) 511–519.

45. J. Setsune, M. Ikeda, Y. Kishimoto and T. Kitao, J. Amer. Chem. Soc., **108** (1986) 1309–1311.

46. R. Cromer, These D'Etat, Univ. of Strasbourg (1986).

47. A.W. Johnson and D. Ward, J. Chem. Soc., Perkin I, (1977) 720–723.

48. H.J. Callot and E. Schaeffer, Nouv. J. Chim., **4** (1980) 307–309.

49. H.J. Callot, Th. Tschamber, B. Chevrier and R. Weiss, Angew. Chem., Int. Ed., **87** (1975) 545.

50. H.J. Callot, B. Chevrier and R. Weiss, J. Am. Chem. Soc., **100** (1978) 4733–4741.

51. H.J. Callot, B. Chevrier and R. Weiss, J. Am. Chem. Soc., **101** (1979) 7729–7730.

52. J.-P. Mahy, P. Battioni and D. Mansuy, J. Amer. Chem. Soc., **108** (1986) 1079–1080.

53. P. Batten, A.L. Hamilton, A. W. Johnson, M. Mahendram, D. Ward and T.J. King, J. Chem. Soc. Perkin I, (1977) 1623–1628.

54. H.J. Callot, Tetrahedron, **35** (1979) 1455–1457.

55. R. Bonnett, R.J. Ridge and E.H. Appleman, <u>J. Chem. Soc.,</u> <u>Chem. Commun.</u>, (1978) 310–311.

56. A.L. Balch, Y.W. Chan, M. Olmstead and M.W. Renner, <u>J. Amer. Chem. Soc.</u>, **107** (1985) 2393–2398.

57. A.H. Jackson, R.K. Panday and E. Roberts, <u>J. Chem. Soc.,</u> <u>Chem. Commun.</u>, (1985) 4700–4701.

58. K.M. Smith and P.K. Pandey, <u>Tetrahedron Lett.</u>, **27** (1986) 2717–2720.

59. J.W. Buckler, In <u>Porphyrins and Metalloporphyrins</u>, K.M. Smith, Ed., Elsevier, NY, NY (1975), pp. 157–231.

60. O.P. Anderson, A.B. Kopelove and D.K. Lavallee, <u>Inorg. Chem.</u>, **19** (1980) 2101–2107.

CHAPTER 6

BIOCHEMISTRY OF N-SUBSTITUTED PORPHYRINS I: FERROCHELATASE INHIBITION

Introduction

The discovery of N-alkylporphyrins in biological systems and the elucidation of the mechanism of their production by the cytochrome P-450 enzymes has been an interesting detective story. It began with the observation by Solomon and Figge that certain drugs cause clinical symptoms in animals that closely resemble those of the porphyric diseases of man.[1,2] These drugs permitted researchers to use animals as models of the human disease. In 1963, Labbe showed that the drug used to induce porphyric symptoms, 3,5-diethoxycarbonyl-1,4-dihydrocollidine, DDC, caused a marked inhibition of ferrochelatase activity.[3] DeMatteis and coworkers discovered that it was not DDC itself, but a green pigment that appeared to be a metabolite of DDC that inhibits ferrochelatase activity.[4] They isolated the green pigment and determined that it had spectral properties like those of porphyrins. The definitive structural determination that the inhibitor was N-methylprotoporphyrin IX was provided by Ortiz de Montellano and coworkers.[5,6]

The obvious manifestation of the inhibition of ferrochelatase, a pronounced increase in iron-free protoporphyrin concentrations in the liver, was crucial to the discovery of N-substituted porphyrins in living systems and to their subsequent use as probes of the structure and mechanisms of the important cytochrome P-450 enzymes. This chapter will be concerned with the use of N-alkylporphyrins as enzyme inhibitors. Following chapters will deal with the N-alkylporphyrins and cytochrome P-450 and the production of N-substituted porphyrins from the interaction of hydrazines with hemoglobin and myoglobin *in vivo* and *in vitro*.

Ferrochelatase, Heme Synthesis and Porphyrias

Ferrochelatase (protoheme ferrolyase or heme synthetase, EC 4.99.1.1) is the terminal enzyme of the heme biosynthetic pathway. Its role is to catalyze the insertion of ferrous ion into protoporphyrin IX to form heme. Enzymes of this class are found in plants, yeasts, bacteria and in the mitochondria of animal cells. Although protoporphyrin IX reacts with ferrous iron at a reasonable rate *in vitro*,[7] liver extracts showing the properties characteristic of enzymes (heat sensitivity, pH optima of reactivity and rates dependent on temperature) catalyze the metal incorporation reaction significantly. In addition, antibodies to the enzyme bovine ferrochelatase have been shown to significantly inhibit the metal incorporation reaction *in vitro*.[8] Reactions which inhibit this enzyme are of interest because they may provide information about the mechanism of formation of heme, which serves as the prosthetic group for a wide variety of cytochromes (electron transfer agents and oxidases) as well as hemoglobin and myoglobin. In addition, hemes are important in the biosynthesis of chlorophylls, but their exact role has been ill-defined. The effects of ferrochelatase inhibitors are of interest as a means of clarifying the mechanism of chlorophyll synthesis. The early studies of ferrochelatase inhibition by N-alkylporphyrins involved extracts of mouse or rat liver microsomes, while more recent studies have employed purified enzyme preparations from bovine or rat liver. In addition, there have been a few studies recently of inhibition of enzyme activity in the photosynthetic bacterium *Rhodopseudomonas spaeroides* and the unicellular rhodophyte *Cyanidium caldarium*. In many studies reported in this chapter, the activity of ferrochelatase was determined by observing the rate at which the extracts or enzyme preparation catalyzed the incorporation of cobaltous ion into mesoporphyrin. Since neither cobaltous ion nor mesoporphyrin are present in significant quantities in the cells from which the enzyme is derived, there is little danger of artifacts from endogenous materials. The molar absorptivities of cobalt mesoporphyrin are high (10^5 in the Soret region near 400 nm and 10^4 in the region of 500 to 700 nm) allowing accurate determination of concentrations as a function of time. A typical value for a rat liver microsome preparation (10% w/v) is 3 nmol/mL per minute or a change of 3×10^{-6} M/min. Purification of extracts results in increases by factors of 1000 to 2000.

In healthy animals, only small quantities of the intermediates on the heme synthesis pathway, δ-aminolevulinic acid and porphobilinogen, are excreted. Porphyrias result in the accumulation of large amounts of these intermediates through increased activity of δ-aminolevulinic acid synthetase. The concomitant production of highly colored porphyrins is sometimes obvious. Porphyrias can be inherited in both man and other animals and porphyric cattle have provided a commercial source of uroporphyrins. In the United States, several thousand individuals suffer from porphyria. In man, the genetically transmitted porphyrias are classified as intermittent, hereditary or variegate. Certain drugs cause severe reactions in porphyric patients. The situation can be serious for patients with mild forms of porphyria whose symptoms may not be evident. Considerable evidence, especially from animal studies, indicates that the enzymatic defect is not in δ-aminolevulinic acid synthetase itself.[9] In variegate porphyria, there is a marked increase in fecal protoporphyrin that suggests that the defect occurs in the conversion of protoporphyrin IX to heme.[10] There is evidence that ferrochelatase is inhibited in skin fibroblasts of porphyric patients in South Africa and the same group has shown that ferrochelatase in normal and porphyric cattle differs by a point mutation that probably causes a modified protein structure.[11] There is evidence, however, that protoporphyrin oxidase but not ferrochelatase is inhibited in American patients with variegate porphyria.[12] Whether or not ferrochelatase is the principal site at which the regulation of heme synthesis is disrupted in the natural porphyrias, it is now evident that drugs which induce the excretion of excessive amounts of porphyrins in animals produce N-substituted porphyrins (green pigments) which inhibit ferrochelatase activity. It is, therefore, reasonable to assume that the inhibition of ferrochelatase by these compounds in the animals limits the amount of free heme to such low levels that either δ-aminolevulinic acid synthetase or protoporphyrin oxidase is deregulated and excess porphyrin production results. It is also reasonable to assume that these "green pigments" can provide valid information about the mode of deregulation, the normal function of free heme, and about the structure and mechanism of action of ferrochelatase itself.

Studies of Inhibition in Liver Microsome Extracts

Because the green pigments that have eventually been identified as N-alkylporphyrins were first noticed as a coloration of the livers of drug-treated animals and because the experimentally induced model porphyria is hepatic, the search for a relationship between the green pigments and inhibition of enzymes in the heme synthesis pathway began with extracts of liver microsomes. Tephly and his coworkers treated mice and rats with DDC (200 mg/kg body weight), then excised the liver after 1h and extracted the liver homogenates twice (50% w/v in 0.25 M sucrose) using an equal volume of 4:1 acetone/ether. To obtain the fraction containing N-alkylporphyrins, the resulting pellet of extracted hemogenate obtained by centrifugation was suspended in 12 mL of acetone and treated with 0.2 mL of concentrated HCl. The extraction was repeated and the extracts combined and evaporated to dryness using N_2. The resulting solid was dissolved in 1 mL of acetone and chromatographed on Sephadex LH-20 columns. Fractions from these columns were evaporated using N_2, dissolved in DMSO and incubated with mitochondrial suspension and pH 8.2 Tris/HCl buffer at 37°C for 15 min. Samples of these solutions were then tested for ferrochelatase activity. A unit of inhibitor was defined based on 50% inhibition of an enzyme activity of 18 nmol of mesoporphyrin converted per min per mL of preincubation mixture. With this procedure, the content of inhibitor (in units of inhibitor/g of liver) was 22.5 for the mouse and 3.0 for the rat.[4] In addition to the spectral similarity of this species to porphyrins, DeMatteis and coworkers demonstrated that the use of δ-amino-^{14}C-levulinate resulted in the incorporation of ^{14}C in the ferrochelatase inhibitor (N-methyl-protoporphyrin IX).[12] The label from 3,5-diethoxycarbonyl-1,4-dihydro(2,6-^{14}C)collidine was not found in the inhibitor.[12]

The inhibition of ferrochelatase activity in mitochondrial preparations has been examined for a number of different drugs that produce liver coloration in rodents. In early studies, when the structures of the green pigments had not yet been defined, it was found that certain drugs such as 2-allyl-2-isopropylacetamide[13], 5-allyl-5(1-methylbutyl)-barbiturate (secobarbitone)[14] and 1-ethynylcyclohexanol[15] produce green livers in rats and mice but the extracted green pigments did not inhibit ferrochelatase activity significantly in the mitochondrial preparations. It was also demonstrated that although DDC results in excess porphyrin production in animals, it is not the intact DDC but rather a metabolite that inhibits ferrochelatase activity in microsomal extracts (DDC itself does not inhibit the formation of cobalt mesoporphyrin).[4] It is now known that the inhibition is due to the production of particular N-substituted porphyrins (metabolites of the drugs that in fact derive most of their structure from the protoporphyrin IX prosthetic group of the cytochrome P-450) and that some of the N-substituted porphyrins ("green pigments") are not inhibitory. De Matteis and Gibbs found that the antifungal drugs griseofulvin and isogriseofulvin act like DDC, producing hepatic green pigments that strongly inhibit ferrochelatase.[16]

In 1980, Tephly and coworkers found that there is a correlation between the amount and rate of inhibitor formed by treatment of rats or mice with DDC and the extent and course of the decrease in mitochondrial ferrochelatase activity in their excised livers.[17] The correlation between the porphyrinogenesis produced by DDC, the inhibition of ferrochelatase and the formation of inhibitor (N–methylprotoporphyrin) established N–methylprotoporphyrin as the substance responsible for experimental animal porphyria produced by DDC. They also tested the activity of oxidized DDC in mice and rats and found that it did not produce the high level of inhibitor (N–methylprotoporphyrin) that results from DDC treatment. What they did find, however, was that in the animals treated with oxidized DDC and control animals that received only oil, only a small amount of the DDC–produced inhibitor was present. The inhibitor present in normal animals was subsequently shown by Tephly and coworkers to be N–methylprotoporphyrin IX.[18] Using nmr spectroscopy, Ortiz de Montellano and coworkers were able to demonstrate that the N–methylprotoporphyrin formed in the livers of DDC treated mice is principally in the same isomeric form as that produced in the livers of DDC treated rats.[19] They showed that the inhibitor present in the livers of untreated mice not only has the same spectrum as this isomer, but has the same TLC R_f value and the same elution time from HPLC (the elution times under their conditions being characteristic of the geometric isomer).[18] Thus, it appears that N–methylprotoporphyrin is present in normal animals and, since it is formed as the same isomer as in DDC treated animals, it is also formed by the same process. Although the procedures necessary for isolation and purification of N–methylprotoporphyrin from untreated mice precluded quantitation, the amounts obtained were stated to be approximately the same as those of free protoporphyrin (data not shown in the article).[18] By pooling liver extracts from 200 mice, they were able to demonstrate that a pigment which resembles in its spectral characteristics the product from DDC treatment also inhibits ferrochelatase to a degree similar to the DDC–produced pigment and authentic N–methylprotoporphyin IX. (13.3, 16.5 and 13.4 inhibitory units per mole, respectively. The authentic N–CH$_3$PP IX is a mixture of isomers.)[20] The authors suggested that, in view of the strong inhibitory properties of N-methylprotoporphyrin, its presence at about the same level as the substrate for ferrochelatase implicates this substance as a possible regulator of ferrochelatase activity. If, in fact, N-alkylporphyrins act as regulators, it would seem that their level should be determined by endogenous substances subject to regulation rather than xenobiotics whose occurrence would be highly variable. Further studies using highly sensitive detection methods will certainly be necessary to resolve the possible role of N-alkylporphyrins in the regulation of heme biosynthesis.

Another item of interest in this article was the finding that the radioactive isotope from ^{14}C-methyl DDC which had been specifically labelled in the 4 position was found in the inhibitor, tracing the origin of the N-methyl group specifically to the 4-position of DDC. They formed a copper complex of from the N-methylprotoporphyrin that was produced from the interaction of [14]-C-methyl DDC with the mouse liver cytochrome P-450 and found that the complex retained the ^{14}C label and was about as strong an inhibitor as N-methylprotoporphyrin itself.

Investigations of the Structural Dependence of Inhibition. When the first phase of structural tests was reported, the nature of the green pigment produced from DDC treatment had been deduced to be N-methylprotoporphyrin IX.[5,21] and it was postulated (later verified) that the product from treatment with ethylene was a two carbon fragment.[14,22] At this point, De Matteis and coworkers investigated the inhibitory properties of N-methylmesoporphyrin IX with the objective of testing the activity of a species derived from a good substrate for the enzyme that has a small N-substituent (the same as N-methylprotoporphyrin IX). They also tested the inhibitory properties of N-ethylmesoporphyrin IX, to determine the relative effect of a two-carbon N-substituent (like that derived from the relatively poor inhibitory pigments produced by ethylene). The third type of N-substituted analog they used was N-methylcoproporphyrin; it provided a test of a derivative of a porphyric product with a small N-substituent, but with a less hydrophobic set of peripheral subsituents. The N-methylcoproporphyrin provided a test of the specificity of ferrochelatase inhibitors that might be related to their ability to mimic the fit of protoporphyrin IX in the heme-binding pocket.[2][3] At the same time, they treated rats with DDC and ethylene and extracted the green pigments from their livers for comparison with the synthetically produced porphyrins of established structure cited above. The major findings shown in Table 6.1 were that both the green pigment from DDC and N-methylmesoporphyrin IX were strongly inhibitory whereas the green pigment (now known to be $N-(CH_2CH_2OH)$protoporphyrin IX) extracted from the livers of animals treated with ethylene , N-ethylmesoporphyrin IX and N-methylcoproporphyrin were much weaker inhibitors. The time of killing after treatment (2 to 22 h) was not a pronounced factor, but there was some difference in results from animals given injections with saline solutions rather than DMSO solutions. It should be borne in mind that there are four structural isomers of chemically produced N-alkylmesoporphyrin (or N-alkylprotoporphyrin) produced in nearly equal amounts by *in vitro* chemical alkylation reactions and each of these exists as a racemic mixture. The isomeric distribution that is produced naturally, however, is not random but is weighted to particular isomers and tautomers. The relative contributions of the isomers present in synthetic and naturally produced N-alkylporphyrins should, therefore, be taken into account when considering the inhibition values that have been reported.

After Ortiz de Montellano and coworkers had successfully separated all four of the structural isomers of N-methylprotoporphyrin IX (the green pigment produced in the livers of DDC treated rats) and established their structure by analysis of their nmr spectra,[6][19] the inhibitory properties of specific isomers could be interpreted.

TABLE 6.1 Inhibition of Ferrochelatase by N-Alkylmesopor-
phyrins and "Green Pigments" from DDC and Ethylene

In Vitro	Inhibitory Activity, units/nm
Green Pigment from DDC treatment (N-methylprotoporphyrin IX)	20.2
N-methylmesoporphyrin IX (synthetic, mixture of isomers)	11.4
N-ethylmesoporphyrin IX (synthetic, mixture of isomers)	0.89
Green Pigment from ethylene (N-2-hydroxyethylprotoporphyrin IX)	0.15
N-methylcoproporphyrin	0.15

Effect of Giving N-alkylporphyrins to Mice

Treatment (and solvent)	Time of Killing (h after treatment)	Ferrochelatase Activity[b]
Control (DMSO)	2	1.46
N-methyl mesoporphyrin	2	0.44
Control (DMSO)	4	1.36
N-methylmesoporphyrin	4	0.47
Control (DMSO)	22	1.38
N-methylmesoporphyrin	22	0.77
Control (saline)	4	1.58
N-methylmesoporphyrin	4	0.48

a Ref. 23.
b nmol of mesoporphyrin utilized per min per mg of enzyme

De Matteis and coworkers separated the geometric isomers of N-methylprotoporphyrin IX and pooled the fractions corresponding to the isomers with the methyl groups on the A and B rings (the N_A and N_B isomers, see Figure 6.1) and those of the C and D isomers (the N_C and N_D isomers) and also tested two of the individual isomers (N_A and N_D), obtaining the results shown in Table 6.2.[1][2] They also studied the inhibitory properties of the zinc complexes of corresponding isomers. Interestingly, there was little difference among the isomers of the free base porphyrin, but a distinctly greater inhibition for the zinc complexes of the N_A and N_B isomers in comparison with the N_C and N_D isomers. A comparison of the structures of an N-methylporphyrin free base[24], and its corresponding zinc complex[25] shows difference in the angle of the N-substituted pyrrole with respect to the mean porphyrin plane of $11°$ ($27.7°$ vs. $38.5°$). A comparison of a monoprotonated N-alkylporphyrin (N-ethoxycarbonylmethyl-octaethylporphyrin)[26] and its cobalt complex[27] shows a much greater difference in this angle, $25°$ ($19.1°$ vs. $44.1°$).

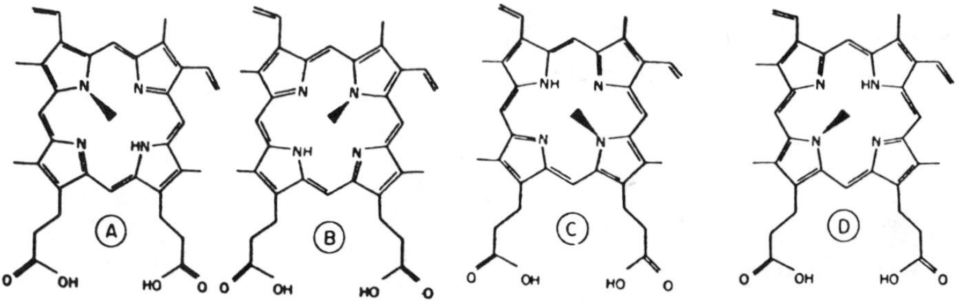

Figure 6.1. The four geometric isomers of protoporphyrin IX. For each of these isomers there are two enantiomers.

Although it is not clear which of these comparisons is most valid for the case of ferrochelatase (the N-alkylproto-porphyrins are in the monoprotonated form at pH 7, but the porphyrin bound in the enzyme may be in the free base form), it is evident that the N-substituted pyrrole ring is tilted from the mean porphyrin plane to a greater extent in the zinc complex than in either the free base or monoprotonated species. The greater cant of the pyrrole ring for the zinc complexes would indeed cause the propionic side chains of the N_C and N_D isomers to be separated to a greater extent than those of the N_A or N_B isomers, as proposed by De Matteis[28]. On one hand, the flexibility of the propionic acid side chains should allow them to adapt to the best position for interaction with the enzyme. On the other, the difference in affinities is only about a factor of six, equivalent to a free energy difference of only 5 kJ/mol, so it could result from less favorable hydrogen bond formation by the N_C and N_D isomers or from even more subtle differences of the fit of the porphyrin to the active site of the enzyme.

TABLE 6.2. Ferrochelatase Inhibition by The Isomers of Free Base and Chlorozinc N-Methylprotoporphyrin IX[a]

Compound Tested	N-methylated Ring	Inhibitory Activity (units/nmol)
N-methylprotoporphyrin	A	10.6
	A and B	12.6
	D	9.6
	C and D	11.7
Chloro-N-methyl-protoporphinatozinc(II)	A	12.8
	A and B	12.9
	D	2.1
	C and D	2.0

a Ref. 28.

The difference in the tilt of the pyrrole ring for a series of zinc complexes does not depend in a simple way on the bulk of the N-substituent (the angles are $38.5°$, $42.0°$, and $32°$ for the N-methyl, N-phenyl and N-benzyl derivatives, respectively),[29] so it does not seem reasonable at this time to attribute the difference in affinities of various N-substituted porphyrins to the different distances between the propionic groups on the porphyrin periphery.

In this same article, De Matteis and coworkers reported the circular dichroism (CD) spectra of N-substituted porphyrins isolated from the livers of rats treated with DDC, ethylene, secobarbitone, 2-allyl-2-isopropylacetamide (AIA) or propyne. The spectra for the two fractions of N-methylprotoporphyrin IX from DDC treated rats had maxima at 430 nm (N_C and N_D) and at 444 nm (N_A and N_B). The species arising from propyne treatment, shown to be the N_A isomer,[30] has its maximum at 445 nm, while that for the N-hydroxy-ethylprotoporphyrin IX derived from ethylene treatment, shown to be the N_C or N_D isomer,[31] has its maximum at 430 nm. Thus, De Matteis concluded that the CD maximum can be used to determine the site of N-substitution. The green pigments from rats treated with AIA and secobarbitone have their maxima at 430 nm, so they were presumed to be substituted predominantly on the C or D rings.

Also in 1981, Ortiz de Montellano and coworkers reported the synthesis and separated (by HPLC) the four regioisomers of N-ethylprotoporphyrin IX, studying the inhibition of rat liver ferrochelatase *in vitro* for each of the isomers.[32] The position of the substitution of the ethyl group was determined by ^{1}H nmr spectrometry, as previously reported for the N-methylprotoporphyrin isomers.[6] They found that the isomers in which the ethyl group is present on pyrrole rings A or B (having vinylic β-substituents) are as strongly inhibitory as the corresponding N-methylprotoporphyrin derivatives (a 50% decrease in enzymatic activity as a ratio of 2.4×10^{-6} g of N-ethylporphyrin to protein), whereas the isomers substituted on rings C or D (which have propionic acid substituents in the α-position) were 30-100 times less inhibitory ($1-3 \times 10^{-4}$ g of N-ethylporphyrin were required to cause a decrease in activity of 50%).[31]

Treatment of rats with the ethyl analog of DCC, 3,5-diethoxycarbonyl-2,6-dimethyl-4-ethyl-1,4-dihydropyridine or DDEP, causes the formation of a green pigment which has been identified as N-ethylprotoporphyrin.[18,19,33,34] Tephly and coworkers found that the N-ethylprotoporphyrin produced was somewhat less inhibitory toward ferrochelatase than is N-methylprotoporphyrin, as later corroborated by DeMatteis and coworkers (see Table 6.3).[35] From the results of Ortiz de Montellano discussed above, this result is reasonable if a mixture of isomers, including some of the two isomers in which the C and D rings are substituted, results from the reaction *in vivo*.

In addition to the inhibitory properties of N-ethyl-protoporphyrin, the means by which its production differs from N-methylprotoporphyrin is of interest. Ortiz de Mon-tellano and coworkers found that rats treated with phenobar-bitol (an inducer of cytochrome P-450) formed more N-ethyl-protoporphyrin upon treatment with DDEP than the amount of N-methylprotoporphyrin formed by DDC treatment of phenol-barbitol pretreated rats.[33] Tephly and his coworkers found a similar result for rats treated with DDEP and DDC without prior treatment with phenobarbitol.[35] In addition, they measured the amount of cytochrome P-450 in the livers of the rats and found that the DDEP destroyed much more (about 3 times, under their conditions) of the cytochrome P-450. The manner in which the cytochrome P-450 is destroyed by the two agents appears to be different and it was suggested that they may act on different members of the class of P-450 enzymes. This subject will be discussed in greater detail in the chapter on N-alkylporphyrins and cytochrome P-450.

The results of the inhibition studies of De Matteis and coworkers were extended and summarized in a 1985 article.[36] In this study, it was shown that the N_C and N_D isomers of N-methyl mesoporphyrin IX, N-ethylmesoporphyrin IX, N-n-pro-pylmesoporphyrin IX, and N-methylprotoporphyrin IX inhibit ferrochelatase activity (as determined by the rate of

cobalt(II) mesoporphyrin formation using rat liver mito-
chrondria preparations) to essentially the same degree (Table 6.3). The first fraction of the isomers of N-methyldeuteroporphyrin which elutes chromatographically (probably
the N_C and N_D isomers) also shows the same activity. Typically, the N_A and N_B isomers are much less inhibitory, with
the pronounced exceptions of N-methyl- and N-ethylprotopor-
phyrins, which show about equal effects, and N-n-propylpro-
toporphyrin IX and the zinc complex of N-methylprotoporphyrin IX, in which the N_A and N_B isomers are significantly
more inhibitory. On the basis of these results, De Matteis
concluded that the interaction of the vinyl groups of protoporphyrin with the enzyme may be quite important and that
bulky N-substituents, like the zinc complexes, show greater
differences between the N_C and N_D isomers and the N_A and N_B
isomers because of greater deviations of the N-substituted
pyrrole ring from the mean porphyrin plane. A comparison of
the structures of N-methyl, N-phenyl and N-benzyl porphyrins
(discussed in Chapter 3), however, shows little difference
in the pyrrole ring - porphyrin plane angle.[29] It seems
reasonable to assume that the bulky N-substitutents are
somewhat weaker inhibitors because they require distortion
of the protein in the region near the center of the
porphyrin substrate.

De Matteis and coworkers have incubated rat hepatocytes
with both AIA and 4-ethylDDC in the presence of exogenous
mesoheme and deuteroheme.[54] The isomers of the N-alkylated
mesoporphyrin were separated by HPLC and shown to be chiral.
The N-alkylprotoporphyrin resulting from the endogenous pool
was also separated into its isomers as discussed above. In
this way, the authors monitored the contributions of
endogenous and exogenous heme. The degree to which the
mesoheme is utilized depends on which P-450 isozyme is
predominant in the hepatocytes.

TABLE 6.3. Ferrochelatase Inhibition of Different Isomers of Several N-Alkylporphyrins[a]

Porphyrin	N-Alkylated Ring	Inhibitory Activity
N-methylprotoporphyrin	A and B	10.8
	C and D	9.24
N-ethylprotoporphyrin	A and B	5.86
	C and D	6.67
N-n-propylprotoporphyrin	A and B	0.11
	C and D	6.06
N-methylmesoporphyrin	A and B	8.01
	C and D	15.0
N-ethylmesoporphyrin	A and B	8.17
	C and D	0.14
N-n-propylmesoporphyrin	A and B	7.85
	C and D	<0.05
N-methyldeuteroporphyrin	A and B (?)	8.78
	C and D (?)	0.32
N-methyluroporphyrin	---	<0.05
N-methylcoproporphyrin	---	0.17
Chloro-N-methylproto-	A and B	15.7
porphinatozinc(II)	C and D	2.1
Chloro-N-mesoporphinato-	A and B	6.96
zinc(II)	C and D	0.5

a Ref. 36.

An interesting observation in this article concerned the inhibition of ferrochelatase by the dimethyl ester of N-methylporphyrin. The reports that this species is inhibitory to nearly the same extent as the corresponding free acid[37,38] were certainly puzzling, since the charge on the periphery of the porphyrin might be expected to be very important in determining how the porphyrin would bind to an enzyme and specifically because the dimethyl esters of porphyrins are inactive as substrates for ferrochelatase. De Matteis demonstrated that while the dimethyl ester of N-methylprotoporphyrin IX is inhibitory toward ferrochelatase activity in mitochondrial preparations, the effect is eliminated by the carboxylesterase inhibitor, bis-(p-nitrophenyl)phosphate. It appears, therefore, that the activity ascribed to the dimethyl ester results from the free acid that is produced by carboxylesterases present in the preparation.

De Matteis and coworkers found that the inhibition caused by N-alkylporphyrins with bulky substituents could be reversed by the addition of substrate (mesoporphyrin IX).[36] This observation could explain why they found N-ethylprotoporphyrin IX isomers N_C and N_D to be strongly inhibitory while Ortiz de Montellano and coworkers, using a different experimental protocol, did not.[32] The N-methylprotoporphyrin IX and N-mesoprotoporphyrin IX species bind so tightly to ferrochelatase that they are difficult to displace with normal substrate and they remain inhibitory. This result does not require that the effect be due to irreversible inactivation – it can still be characterized as competitive.

Ferrochelatase inhibition has also been studied in chick embryo[39] and chick embryo liver cell culture[40,41] in order to determine which feature of the structure of DDC is crucial to its inhibitory activity. Cole and coworkers found that an alkyl group at position 4 is essential, corroborating De Matteis's finding that a ^{14}C label on methyl groups at the 2 and 6 positions are not transferred to protoporphyrin IX.[39,40] They also found that active DDC analogues have ester groups at positions 3 and 5 and are dihydropyridines.

Marks and coworkers studied the ferrochelatase-reducing activity and cytochrome P-450 and heme-destructive effects of a number of analogs of DCC in chick embryo liver cells.[41] They found three types of results: some derivatives of DDC destroy cytochrome P-450, cause accumulation of protoporphyrin, and produce metabolites that inhibit ferrochelatase (such as DCC itself, and the 4-ethyl and 4-propyl derivatives, as also found by Ortiz de Montellano[42]), a second group neither destroys cytochrome P-450 nor results in ferrochelatase inhibition (for example, dimethyl-DDC, which has no 4-methyl group to transfer, oxidized DDC, and 4-phenyl DDC), and a third group causes cytochrome P-450 destruction but results in the accumulation of copro- uro and/or heptacarboxylporphyrins rather than protoporphyrin and does not produce metabolites that inhibit ferrochelatase (the 4-benzyl, 4-isopropyl, 4-cyclohexyl and 4-(3-cyclohexenyl) derivatives). Aspects of this chemistry will be discussed in detail in a subsequent chapter on cytochrome P-450 reactions, but the important point of this study in relation to those discussed above is the common feature of the inhibition of the active DDC analogues: they produce N-alkylporphyrins (with the N-ethyl group arising from the 4-alkyl moiety of the DDC analog).[18, 34, 40, 43]

Studies Employing Purified Enzyme

Although the existence of catalytic iron chelating activity ascribed to an enzyme (ferrochelatase) has been recognized and examined for many years, the actual isolation and purification of the enzyme was not accomplished until recently. In earlier attempts, nearly all the catalytic activity was lost. Ferrochelatase is tightly bound to the inner mitochondrial membrane in eukaryotes and to the cytoplasmic membrane of bacteria, requiring use of detergent. In later stages of the procedures when the detergent was removed, the activity of the enzyme was lost. In 1980, Mailer, *et al.* purified ferrochelatase from rat liver using 2-vinyl-4-(3'-

(N-3''-aminopropyl)-acrylamidodeuteroporphyrin on an affinity column and found that 2-mercaptoethanol was required to
preserve activity (and, at that, for only a 12 h half-life
at 4°).[44]   Blue Sepharose CL-6B chromatography has been used
to purify the enzyme from rat liver[45], bovine liver,[8,46]
and the bacterium *Rhodopseudomonas sphaeroides.*[47]   The purified rat liver enzyme contains considerable amounts of fatty
acids and is enhanced markedly by addition of exogenous
lipid, whereas the purified bovine liver enzyme contains
neither fatty acids nor phospholipids and is enhanced much
less by addition of lipid.[9]   Ferrochelatase does not appear
to require cofactors.   Thus, if care is taken to use the
same amount and type of lipid in the test medium, results
from different preparations of ferrochelatase should produce
comparable results.

Dailey and Fleming chose to work with bovine ferrochelatase because a form of genetic porphyria resembling the
disease in man has been described for cattle.[46]   They found
that N-methylprotoporphyrin IX (a mixture of isomers from a
synthetic preparation which De Matteis had shown are inhibitory to about the same extent) at low concentrations does
not irreversibly titrate the enzyme (Figure 6.2). They determined a K_i value for N-methylprotoporphyrin IX of 7 nM.
Their caveat against applying the procedures they employed
with the purified enzyme in studies of crude extracts (because of the lipophilic nature of the porphyrins and the
possible binding of porphyrins to other enzymes) is well
taken.   N-methylprotoporphyrin IX was found to be a noncompetitive inhibitor with respect to iron and a competitive
inhibitor with respect to non-N-substituted porphyrin substrates, supporting an ordered sequential enzyme mechanism
in which iron is bound before the porphyrin (Scheme 6.1).

Scheme 6.1 was further supported by their data for inhibition by manganese, which is competitive with respect to iron and noncompetitive with respect to porphyrin. This study represents the first thorough kinetic analysis of ferrochelatase inhibition carried out with purified enzyme and the data provide clear support for their conclusions. The implications of the inhibition studies of ferrochelatase from a wide variety of sources using different metal ions as well as active-site-structure probes will be discussed in detail in a review of the field.[55]

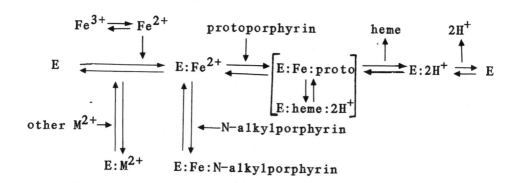

Scheme 6.1

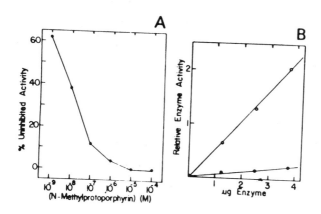

Figure 6.2. Plots demonstrating that N-methylprotoporphyrin IX binds reversibly to bovine ferrochelatase.[46]

Hemeproteins are degraded in animals to produce bile pigments which apparently have no function except as a means of eliminating unneeded heme. In plants a blue-green algae, however, the bilins with similar structures to the biliverdin and bilirubin produced in animals play functional roles in photosynthesis and photomorphogenesis. The photosynthetic antennae of plants and some algae are composed of bilins associated with proteins. Troxler, Brown and their coworkers found that there is a precursor-product relationship between the plant bilins and heme.[48-50] Since N-methylprotoporphyrin has been shown to inhibit ferrochelatase in mammalian systems, Brown and coworkers reasoned that it could be used to determine the role of ferrochelatase in the plant biosynthesis of bilins.[37] They chose the unicellular rhododyte *Cyanidium caldarium* because it shows ferrochelatase activity and because it can be grown aerobically in the dark without the formation of photosynthetic pigments and then induced to form photosynthetic pigments in two quite different ways: by suspension in minimal medium and exposure to light, giving phycocyanin and chlorophyll a without significant cell division; and by resuspension in the dark in a medium containing glucose and δ-aminolevulinate, resulting in excretion of large amounts of porphyrins and phycocyanobilin, the bilin chromophore of phycocyanin.[51] They studied the effect of N-methylprotoporphyrin on phycocyanobilin and chlorophyll a synthesis by measuring the absorption of light by aliquots of the supernatant from centrifuged cell preparations at wavelengths charactersitic of various species (610 nm for phycobilin, 675 nm for chlorophyll a, and 620 nm for phycocyanin).

Brown and coworkers found that both N-methylprotoporphyrin and the dimethyl ester strongly inhibit bilin synthesis both in the light and in the experiment in the dark where 5-aminolevulinic acid is added to promote the synthesis of phycocyanobilin.[37] This result implies that ferrochelatase (and, hence, heme) is required for algal bilin synthesis. It is curious that the dimethyl ester is active in this

case, but, again, it may be that it is deesterified before
it inhibits the ferrochelatase.

They also found that N-methylprotoporphyrin has little
effect on the synthesis of chlorophyll a in the dark (even
the step in which magnesium is inserted is not inhibited in
the dark), however, the synthesis of chlorophyll a in the
light is inhibited in a manner parallel to the inhibition of
phycocyanin synthesis. The explanation they have advanced
for these results is that the primary inhibition of chlorophyll a synthesis in the light is actually the inhibition of
phycocyanin synthesis and that the synthesis of chlorophyll
in the light is strictly coordinated to phycocyanin
synthesis. The results also imply that the synthesis of
chlorophyll a in this organism occurs by different routes in
the dark and in the light.

Houghton and coworkers studied inhibition of ferrochelatase by N-methylprotoporphyrin dimethyl ester in isolated
membranes and growing cultures of the photosynthetic
bacterium *Rhodopseudomonas sphaeroides* in order to determine
the relationship between heme and bacteriochlorophyll
synthesis in this organism.[38] Under conditions of high
aeration, this organism is able to utilize nutrients in the
medium without assembling the photosynthetic apparatus
characteristic of the photosynthetic form. In this case,
hemes that serve as prosthetic groups of cytochromes are
synthesized, but the magnesium tetrapyrroles used for photosynthesis are not. There is evidence that both the hemes
Ω and the bacteriochlorophylls originate from protoporphyrin,[52] but under conditions of iron deficiency, coproporphyrin is the final product rather than the magnesium pigments.

N-methylprotoporphyrin dimethyl ester was found to be a
potent inhibitor of ferrochelatase in *Rps. sphaeroides* membranes (30% inhibition was measured at a concentration of
1.5 nanomolar). A lag time of a few seconds was noted (the
time for deesterification, perhaps). In cultures of whole
cells of *Rps. sphaeroides*, N-methylprotoporphyrin dimethyl
ester at 8×10^{-7} M caused a decrease in cytochrome b and c
content by 40% and the ferrochelatase activity is reduced by

a factor of 10. The inhibitor-induced decrease in the concentration of free heme in the cells is accompanied by the production of magnesium tetrapyrroles. This result indicates that the N-methylprotoporphyrin is specific for enzymatic iron chelation and does not affect magnesium chelation, consistent with the results found by Brown and coworkers for *Cyanidium caldarium*[38], as discussed above. In cells of *Rps. sphaeroides* competent to make bacteriochlorophyll, inhibition of ferrochelatase by N-methylprotoporphyrin causes the same effects on pigment production as are found under conditions of iron deficiency. In green plants, similar effects are found under conditions of iron deficiency,[53] implying that the synthesis of chlorophyll in bacteria and higher plants may be quite similar.[38] The point was made that the normal ferrochelatase activity in these cells is far in excess of that needed for the synthesis of cytochromes and, hence, sufficient tetrapyrroles could be made even in the presence of inhibitors.

Inhibition of Heme Oxygenase by N-Alkylporphyrins

Heme oxygenase is the rate limiting enzyme in heme degradation. This enzyme can accept dicarboxylic porphyrin complexes other than heme as substrates, generally decomposing iron complexes into bile-like pigments and being competitively inhibited by zinc complexes. De Matteis and his coworkers found that zinc complexes of N-alkylporphyrins inhibit heme oxygenase,[36] as shown in Table 6.4. In general, the relative ability of an N-alkylporphyrin to inhibit heme oxygenase parallels its ability to inhibit ferrochelatase. The levels of N-alkylporphyrin complex needed (at least in rat liver mitochondrial preparations *in vitro*), however, are much higher than those needed to inhibit ferrochelatase. It is possible that at least some of the disruption of the heme biosynthetic pathway caused by N-alkylporphyrins could be due to inhibition of heme oxygenase, which could allow protoporphyrin concentrations

to rise to the point that heme itself inhibits ferrochelatase. However, it has not been demonstrated that the free
base N-alkylporphyrins inhibit the enzyme, there is no
evidence for the presence of metal complexes of N-alkylporphyrins *in vivo*, and, finally, the larger concentrations
required for inhibition *in vitro* suggest that ferrochelatase
is the more likely target for inhibition. It is, of course,
possible that the concentrations of N-alkylporphyrin in the
vicinity of heme oxygenase *in vivo* could be effectively
increased by favorable transport or intermolecular associa-
tion processes, but no evidence for such specific mechanisms
now exists.

TABLE 6.4 Inhibition of Heme Oxygenase from Rat Liver by
Zinc Complexes of N-Alkylporphyrins[a]

Zinc Complex	% Inhibition of Heme Oxygenase Inhibitor Conc., $M \times 10^7$		
	2.5	4.7	15
N-methylprotopor- phyrin (N_A isomer)	56	81	--
N-methylprotopor- phyrin (N_D isomer)	25	42	--
N-n-propylprotopor- phyrin ($N_A + N_B$)	31	66	--
N-n-propylprotopor- phyrin ($N_C + N_D$)	--	6	42

[a] Ref. 36.

References

1. H.M. Solomon and F.H.J. Figge, Proc. Soc. Exp. Biol. Chem., **100** (1959) 583-586.

2. F. DeMatteis, Pharmacol. Rev., **19** (1967) 523-557.

3. J. Onisawa, R.F. Labbe, J. Biol. Chem., **238** (1963) 724-727.

4. T.R. Tephly, A.H. Gibbs and F. DeMatteis, Biochem. J., **180** (1979) 241-244.

5. P.R. Ortiz de Montellano, K.L. Kunze and B.A. Mico, Molec. Pharmacol., **18** (1980) 602-605.

6. K.L. Kunze and P.R. Ortiz de Montellano, J. Am. Chem. Soc., **103** (1981) 4225-4230.

7. R.J. Kassner and H. Walchak, Biochim. Biophys. Acta, **304** (1973) 294-303.

8. S. Taketani and R. Tokunaga, Eur. J. Biochem., **127** (1982) 443-447.

9. S.P.C. Cole and G.S. Marks, Molec. and Cell. Biol., **64** (1984) 127-137.

10. D.J. Viljoen, E. Cayanis, D.M. Becker, S. Kramer, B. Dawson and R. Bernstein, Am. J. Hematol., **6** (1979) 185-190.

11. a) D.A. Brenner and J.R. Bloomer, N. Eng. J. Med., **302** (1980) 765-769.
    b) J.R. Bloomer, H.D. Hill, K.O. Morton, L.A. Anderson-Burnham and J.G. Straka, J. Biol. Chem., **262** (1987) 667-671.

12. F. DeMatteis, A.H. Gibbs and T.R. Tephly, Biochem. J., **188** (1980) 145-152.

13. A.F. McDonaoh, R. Pospisil and U.A. Meyer, Biochem. Soc. Trans., **4** (1976) 297-298.

14. F. De Matteis and L. Cantoni, Biochem. J., **183** (1979) 99-103.

15. P.R. Ortiz de Montellano, K.L. Kunze, G.S. Yost and B.A. Mico, Proc. Natl. Acad. Sci., U.S.A., **76** (1979) 746-749.

16. F. DeMatteis and A.H. Gibbs, Biochem. J., **146** (1975) 285-287.

17. F. De Matteis, A.H. Gibbs and T.R. Tephly, <u>Biochem. J.</u>, **188** (1980) 145-152.

18. T.R. Tephly, B.L. Coffman, G. Ingall, M.S. Abou Zeit-Har, H.M. Goff, H.D. Tabba and K.M. Smith, <u>Arch. Biochem. and Biophys.</u>, **212** (1981) 120-126.

19. P.R. Ortiz de Montellano, P.R. Beilan and K.L. Kunze, <u>Proc. Natl. Acad. Sci., U.S.A.</u>, **78** (1981) 1490-1494.

20. F. DeMatteis, A.H. Gibbs, P.B. Farmer, J.H. Lamb and C. Hollands, <u>Adv. Pharmacol. Ther. Proc., 8th Int. Conf.</u>, Pergamon Press, New York, **5** (1982) 131-138.

21. F. De Matteis, and A.H. Gibbs, <u>Biochem. J.</u>, **187** (1980) 285-288.

22. P.R. Ortiz de Montellano, K.L. Kunze and G.S. Yost, <u>Biochem. Biophys. Res. Comm.</u>, **83** (1978) 132-137.

23. F. De Matteis, A.H. Gibbs and A.G. Smith, <u>Biochem. J.</u>, **189** (1980) 645-648.

24. D.K. Lavallee and O.P. Anderson, <u>J. Amer. Chem. Soc.</u>, **104** (1982) 4707-4708.

25. D.K. Lavallee, O.P. Anderson and A. Kopelove, <u>J. Am. Chem. Soc.</u>, **100** (1978) 3025-3033.

26. G.M. McLaughlin, <u>J. Chem. Soc., Perkin II</u>, (1974) 136-140.

27. D.E. Goldberg and K.M. Thomas, <u>J. Am. Chem. Soc.</u>, **98** (1976) 913-917.

28. F. De Matteis, A.H. Jackson, A.H. Gibbs, K.R.N. Rao, J. Atton, S. Weerasinghe and C. Hollands, <u>FEBS Lett.</u>, **142** (1982) 44-48.

29. C.S. Sauer, O.P. Anderson, D.K. Lavallee, J.-P. Battioni and D. Mansuy, <u>J. Amer. Chem. Soc.</u>, **109** (1987).

30. P.R. Ortiz de Montellano and K.L. Kunze, <u>Biochem.</u>, **20** (1981) 7266-7271.

31. P.R. Ortiz de Montellano, H.S. Beilan, K.L. Kunze and B.A. Mico, <u>J. Biol. Chem.</u>, **256** (1981) 4395-4399.

32. P.R. Ortiz de Montellano, K.L. Kunze, S.P.C. Cole and G.S. Marks, <u>Biochem. Biophys. Res. Comm.</u>, **103** (1981) 581-586.

33. P.R. Ortiz de Montellano, H.S. Beilan and K.L. Kunze, <u>J. Biol. Chem.</u>, **256** (1981) 6708-6710.

34. F. De Matteis, A.H. Gibbs, P.B. Farmer and J.H Lamb,
    FEBS Lett., **129** (1981) 328-331.

35. B.L. Coffman, G. Ingall and T.R. Tephly, Arch. Biochem.
    Biophys., **218** (1982) 220-224.

36. F. DeMatteis, A.H. Gibbs and C. Harvey, Biochem. J., **226**
    (1985) 537-544.

37. S.B. Brown, J.A. Holroyd, D.I. Vernon, R.F. Troxler and
    K.M. Smith, Biochem. J., **208** (1982) 487-491.

38. J.D. Houghton, C.L. Honeybourne, K.M. Smith, H.D. Tabba
    and O.T.G. Jones, Biochem. J., **208** (1982) 479-486.

39. S.P.C. Cole and G.S. Marks, Int. J. Biochem., **12** (1980)
    989-992.

40. S.P.C. Cole, R.A. Whitney and G.S. Marks, Mol.
    Pharmacol., **20** (1981) 395-403.

41. G.S. Marks, D.T. Allen, C.T. Johnston, E.P. Sutherland,
    K. Nakatsu, and R.A. Whitney, Mol. Pharmacol., **27**
    (1985) 459-465

42. O. Augusto, H.S. Beilan and P.R. Ortiz de Montellano, J.
    Bio. Chem., **257** (1982) 11288-11295.

43. P.R. Ortiz de Montellano, H.S. Beilan and K.L. Kunze, J.
    Bio. Chem., **10** (1981) 6708-6713.

44. K. Mailer, R. Poulson, D. Dolphin and A.D. Hamilton,
    Biochem. Biophys. Res. Comm., **96** (1980) 777-784.

45. S. Taketani and R. Tokunaga, J. Biol. Chem., **256** (1982)
    12748-12753.

46. H.A. Dailey and J.E. Fleming, J. Biol. Chem., **258** (1983)
    11453-11459.

47. H. Dailey, J. Biol. Chem., **257** (1982) 14714-14718.

48. R.F. Troxler and S.B. Brown, Biol. Bull., **159** (1980)
    502.

49. S.B. Brown, J.A. Holroyd, R.F. Troxler and G.D. Offner,
    Biochem. J., **6** (1969) 116-132.

50. S.B. Brown, J.A. Holroyd, R.F. Troxler and G.D. Offner,
    Biochem J., **194** (1981) 137-147.

51. R.F. Troxler and L. Bogorad, Plant Physiol., **41** (1966)
    491-499.

52. O.T.G. Jones, in The Photosynthetic Bacteria, R.K. Clayton and W.R. Sistrom, eds. Plenum Press, New York (1978) pp. 750-777.

53. S.C. Spiller, A.M. Castelfranco and P.A. Castelfranco, Plant. Physiol., 69 (1982) 107-111.

54. F. De Matteis, C. Harvey and S.R. Martin, Biochem. J., 238 (1986) 263-268.

55. D.K. Lavallee, in Molecular Structure and Energetics, Vol 9, Liebman, J. and Greenberg, A., eds., VCH Publishers, Inc., New York, N.Y., 1988.

CHAPTER 7

BIOCHEMISTRY OF N-SUBSTITUTED PORPHYRINS II: FORMATION BY CYTOCHROME P-450

Introduction

Considerable time has passed since the discovery that certain drugs which produced symptoms of hepatic porphyria also cause the formation of "green pigments" in the livers of drug-treated animals (as discussed in the last chapter) and, further, that there is a concomitant loss of cytochrome P-450 activity. Only recently, however, was it been demonstrated that the phenomenon causing these results is the destruction of cytochrome P-450 enzymes by its substrates to form N-substituted porphyrins. The early work, before the establishment of the basic mechanism of action of these drugs on cytochrome P-450 enzymes, has been reviewed by DeMatteis[1,2] and Ortiz de Montellano.[3] More recent reviews, which summarize the effects of a wide variety of drugs have been authored by Ortiz de Montellano.[4,5]

With respect to the N-substituted porphyrins, suicide substrate reactions are of interest from two aspects: the mechanism of the production of N-substituted porphyrins and the subsequent interactions of these N-substituted porphyrins with other biomolecules. An important feature of N-substituted porphyrin production from the cytochrome P-450 enzymes is the nature of the precursors: the range of types of functional groups or combinations of functional groups that are required for this reaction. This chapter will deal with these mechanistic and structural aspects, including the contributions of model studies to our understanding of the biological processes. The second area of interest is the biochemical and sometimes clinical results caused by the presence of N-substituted porphyrins, as discussed in the previous chapter for ferrochelatase and heme oxygenase.

The Biochemical Roles of Cytochrome P-450 Enzymes

The cytochrome P-450 enzymes are a class of monooxygenases which contain iron protoporphyrin IX at their active site. Upon reduction of the ferric resting state to the ferrous state, these enzymes readily bind carbon monoxide, giving rise to a shift of the intense Soret band to the unusual position of about 450 nm. The absorbance at 450 nm is not a feature of the functional natural system. The absorption maxima of these "P-450 enzymes" under normal conditions in which the enzymes is active are very much like those of other proteins containing iron porphyrins and it is difficult to detect these enzymes without the reduction procedure. These enzymes are found in the hepatic endoplasmic reticulum, the adrenal cortico mitochondria, renal brush border and the plasma membranes of bacteria. While some cytochrome P-450 isozymes have very specific functions, such as the production of hydroxysteroids in the adrenal cortex, the function of interest with respect to N-alkyl porphyrin production is the more general function of the hepatic enzymes to catalyze the hydroxylation of lipophilic foreign substances, producing derivatives that are more readily excreted. The detoxification ability of the cytochrome P-450 enzymes is highly diverse: this is not a class of enzymes limited to one or a few specific substrates. The types of metabolites that can be oxidized include the $\omega-CH_3$ group of an alkane or fatty acid, the ring carbons of a steroid or camphor, polycyclic aromatic hydrocarbons such as benzpyrene or methylcholanthrene, phenobarbitol, and the N-alkyl substituents of a variety of secondary and tertiary amines. In addition to hydroxylation, reactions with these substrates can produce epoxides by addition across a double bond (in some cases, giving the terminal product) or by simple addition of an oxygen atom to a non-bonding electron pair of the substrate. There is considerable evidence that there are several forms of cytochrome P-450 present within a given organism (or, for example, within the liver itself in higher animals). In somes cases, there is direct spectral

evidence for different isozymes: exposure of a reduced
mitochondrial preparation gives a maximum absorbance
slightly shifted from 450 nm, for exmaple, at 448 nm or 452
nm. In some cases, these distinct isozymes are inducible by
specific zenobiotic reagents. The existence of several
isozymes strongly indicates that there each one is limited
to a particular range of substrates.

The cytochrome P-450 enzymes are not independent of other
biochemical processes, but are, in fact, regulated by a
variety of mechanisms and are themselves regulators of such
important, highly-regulated processes as hepatic heme biosynthesis. Therefore, a change in the activity of these
enzymes is of interest not only for the effect that it has
on the ability of the liver to detoxify xenobiotics, but
also for the corollary effects that it has on other essential processes. The cytochrome P-450 enzymes can be inhibited in several quite different ways that are related only
in a very general sense to the suicide substrate inhibition
which produces N-substituted porphyrins. Reviews of several
of these alternative processes have been published[6-10] and
this chapter will be concerned solely with processes involving N-substituted porphyrins.

An experimentally important feature of the biochemistry
of the cytochrome P-450 enzymes is that they are inducible.
The amount of enzyme increases by over an order of magnitude
when bacteria (*Pseudomonas putida* being commonly employed)
are treated with camphor or when animals are treated with a
variety of xenobiotics, notably DDT, methylchloanthrene or
phenobarbitol. For studies employing microsomal preparations, rats are commonly treated with phenolbarbitol before
their livers are removed for isolation of the microsomal
fraction.

The presence of cysteine in the sequences of a number of
cytochrome P-450 enzymes, the correlation of the long wave-
length absorption of the ferrous state in the presence of
carbon monoxide and EPR and EXAFS spectral properties of the
enzyme led a number of investigators to propose that the
fifth ligand to the iron is a cysteinyl sulfur, with the

sixth site accessible to a large hydrophobic pocket. This hypothesis has been verified by the crystal structure for cytochrome P-450 from *Pseudomonas putida* (with camphor bound at the active site) reported by Poulos and coworkers.[11]   In spite of the significant accomplishment that this structural determination by x-ray crystallography represents, spectral data will continue to be very important for making structural comparisons since it is unlikely that a large number of crystallographic determinations for such difficult systems will be performed.

A catalytic cycle for the normal reactions of hepatic cytochrome P-450 enzymes (reprinted with permission of *Pure and Appl. Chem.*) is presented in Scheme 7.1. The nature of the highly oxidized intermediate which is present at the step where hydroxylation occurs has been the subject of intense research with model compounds and has led to a range of new catalytic oxidation reactions of metalloporphyrins that may find practical applications.

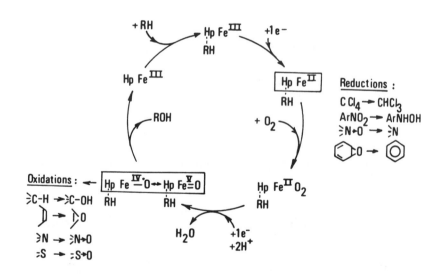

Scheme 7.1

N-Alkylation of Cytochrome P-450 Enzymes

It is at the step in the catalytic sequence in which the highly oxidized iron center has been formed that production of N-alkylporphyrins is thought to occur. Irreversible N-alkylation initially produces a complex which can be assumed from the chemical properties of N-alkylporphyrin complexes of iron to be catalytically inactive (they have greatly shifted reduction potentials favoring the reduced state, making it difficult to return to the ferric resting state or produce further oxidized intermediates[12,13]). Eventually, the N-alkylprotoporphyrin is lost from the active site. The N-alkylporphyrins have been *isolated* as a free base (or monoprotonated species) without the iron atom.

Currently, it is not clear whether it is the iron complex or the free base that is *initially extruded from the protein*. De Matteis and coworkers have presented evidence that [59]Fe (injected 17h before 2-isopropyl-4-penteneamide) is present in the green fraction (which gives the characteristic visible-uv spectrum of an N-substituted protoporphyrin after acidic demetallation and neutralization) separated by chromatography on Sephadex LH-20 using 5% acetic acid in acetone (the second band, protoporphyrin IX, was removed with 10% acetic acid).[14] Both bands were radioactive and showed a strong absorption at 400 nm. Treatment with AIA caused a loss of absorbance and of radioactivity in the slow-moving band and a concomitant increase in the first, green band. Thus, the conclusion was drawn that the iron was still bound to the N-alkylprotoporphyrin. However, we have found that iron is very easily removed from N-alkylprotoporphyrin and, indeed, by 5% acetic acid in acetone. Also, the visible-uv spectrum presented for the green band[14] is much more similar to that of the monoprotonated N-alkylprotoporphyrins than to the iron complexes of N-alkylprotoporphyrin IX, which have an absorbance in the Soret region nearly 30 nm to the red. We have noticed that non-covalently-bound metal ions tend to migrate on chromatographic columns along with anionic porphyrins, so it is possible that the iron is not actually covalently bound. It is surprising, however, that no fluorescence was observed for the green band. The monoprotonated form of N-alkylporphyrins fluoresces less than the free base, but it is still significant. Possibly the fluoresence was quenched by an iron-containing anion in close proximity (external heavy-atom effect) or there was sufficient ferric chloride present to absorb the light near 400 nm. Neither of these possibilities is especially appealing, however, and this question deserves further study.

In a different type of study with a related conclusion, White found that green pigments are produced in the bile of rats treated with the contraceptive steroid norethindrone.[15a]. These green pigments are almost certainly N-substituted porphyrins. The amount of green pigments in the bile was related to destruction of cytochrome P-450 (being reduced in rats treated with cycloheximide) and they show a strong absorption in the Soret region, like porphyrins. In addition, mass spectral data for green pigments formed in livers of norethindrone-treated rats are consistent with structures of N-substituted porphyrins.[15b]. He found that rats dosed with ^{59}Fe as $FeCl_3$ prior to treatment with norethindrone gave bilary extracts with significantly higher ^{59}Fe activity than extracts of rats treated only with the ^{59}Fe. TLC of the green pigments from the bile of the norethindrone-treated rats, esterified under neutral conditions, showed ^{59}Fe activity and did not fluoresce red under uv light. Thus, he concluded that iron was still bound. Several factors mitigate against this conclusion, however. First, the absorption maximum of the extracts was 417 nm, closer to that of the monoprotonated N-alkylprotoporphyrins than that of the iron complex. Secondly, although the esterification of the extracts was performed under neutral pH conditions, a pH of 2.0 (adjusted using 4M HCl) was used for the extraction step preceding the esterification. The iron atom would certainly have been removed from an N-substituted porphyrin at that pH. The fact that the iron could not be found after this extraction and, therefore, during the TLC experiment, demonstrates that the lack of fluorescence and/or presence of ^{59}Fe activity in a chromato-graphic fraction is not conclusive proof for the presence of covalently-bound iron.

General Mechanisms of N-Alkylporphyrin Production From Cytochrome P-450 Enzymes

Terminal Olefins. Ortiz de Montellano has proposed a general mechanism in which N-alkylporphyrins are produced from terminal olefins in a process that is competitive with epoxide formation, as shown in Scheme 7.2.[16] This mechanism is illustrated in Scheme 7.3 for the specific case of AIA, 2-isopropyl-4-penteneamide (commonly called allylisopropyl-acetamide), which produces an N-alkylated product in about one of every 200 turnovers.[5] From the manner in which changes in reaction conditions produce different ratios of epoxide product to N-alkylprotporphyrin (discussed later in the chapter), Ortiz de Montellano has proposed the more detailed mechanism shown in Scheme 7.4. In this scheme, there are three routes by which an alkene could be converted to an N-alkylporphyrin (by a carbonium ion complexed to iron(III), by a radical complexed to iron(IV) or by prior formation of an uncomplexed cation radical) and the possible routes of interconversion of initial and subsequent interme-diates is illustrated.

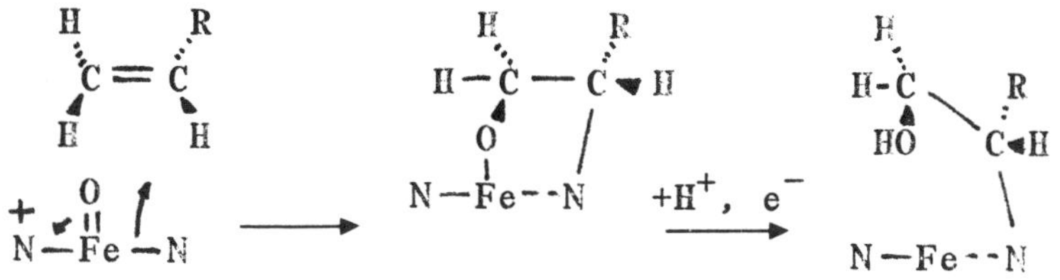

Scheme 7.2

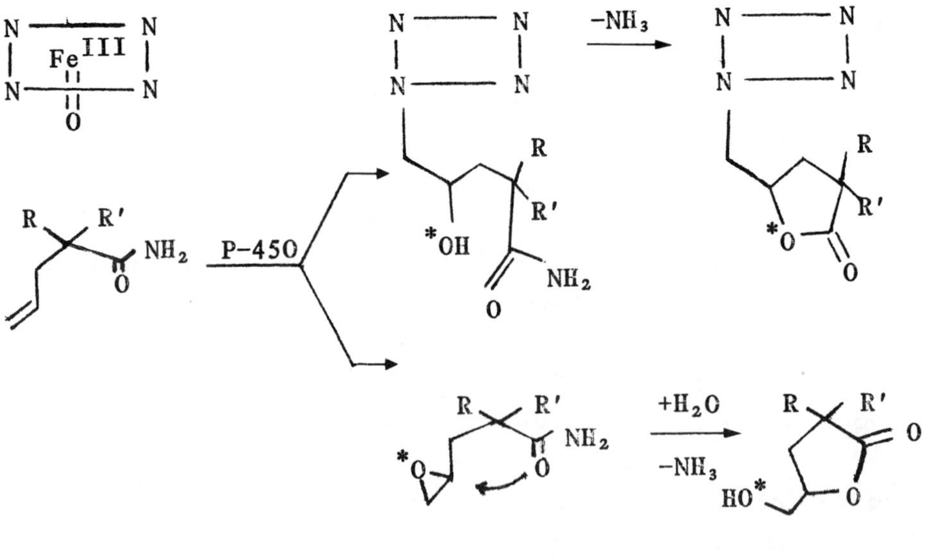

Scheme 7.3

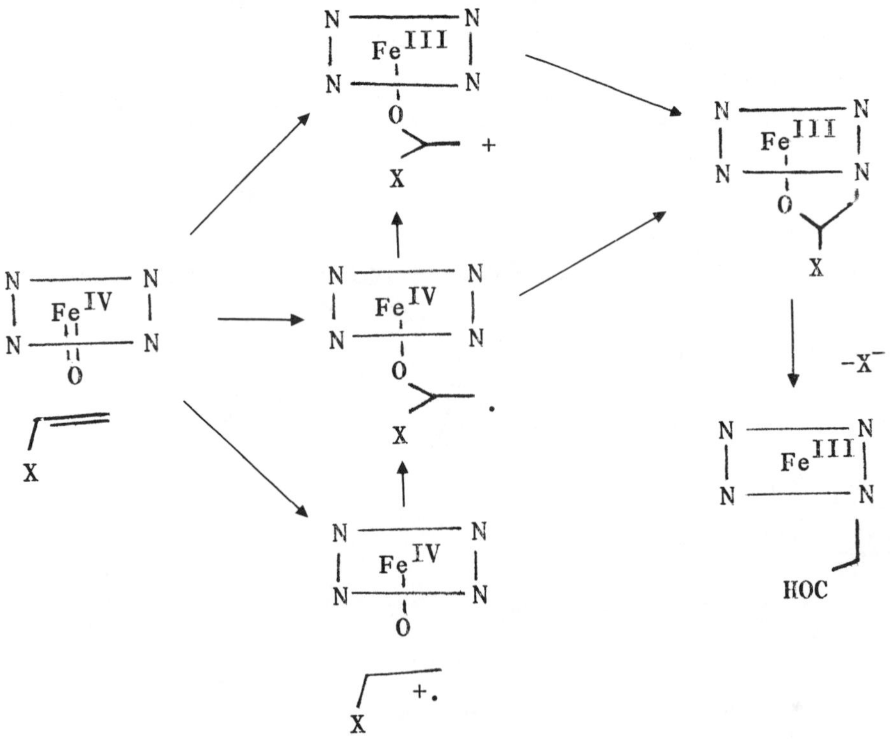

Scheme 7.4.

Considerable evidence from the reactions of cytochrome P-450 enzymes (to be discussed below) supports the general mechanism with the complexed carbonium ion and iron(III) as the principle path. In addition, the groups of Dolphin, Traylor[17] and Mansuy[18] have isolated metallocyclic intermediates from the reactions of synthetic iron porphyrins, strong oxidants and terminal olefins. These provide corollary support for a mechanism involving both the interaction of the oxygen atom bound to the iron and formation of the bond between a pyrrolidinenic nitrogen and the terminal carbon atom at the same time. The reaction sequence proposed by Dolphin, Traylor and coworkers, as well as a discussion of the development and interpretation of data for such models, are presented at the end of this chapter.

AIA and other allyl-barbiturates. The substrate AIA is of historical importance in the development of the biochemical mechanism for suicide substrate inhibition of cytochrome P-450 enzymes. Wade and coworkers in 1968[19] and Waterman and coworkers in 1969[20] reported that the concentration of cytochrome P-450 in the livers of mice and rabbits, respectively, decreased upon administration of AIA. To determine whether the effect of AIA was on the synthesis of cytochrome P-450 or on the destruction of existing cytochrome P-450, De Matteis and coworkers performed several experiments. They administered AIA to animals having elevated P-450 levels (by pretreatment with phenolbarbitone) and found that the decrease in P-450 was much faster than after inhibition of protein synthesis by cycloheximide.[21,22] They also determined that the loss of cytochrome P-450 was accompanied by loss of microsomal heme but that both the amounts of cytochrome b_5 and the protein content were unaltered.[21] These findings demonstrated that the AIA caused these effect by destroying existing heme and not by producing a general alteration of the the endoplasmic reticulum. By using ^{3}H and ^{14}C labelled heme, they and other workers found that AIA treatment led to reduced radioactivity of the crystalline hemin isolated from liver microsomes, showing that AIA treatment alters the heme of cytochrome P-450.[23-25]

The first clue to the discovery that the effects of AIA are due to the production of N-substituted porphyrins was provided by the observation by Schwartz and Ikeda in 1955 that rats and rabbits treated with AIA or the related drug allylisopropylacetylurea had discolored livers.[26] They extracted the livers and found that the discoloration was due to a compound with very strong absorption in the same region as porphyrins and they suggested that these might be porphyrin-like intermediates in the biosynthetic pathway to heme. However, DeMatteis and coworkers found that there is a direct correlation of the loss of cytochrome P-450 and the degree of brownish-green liver coloration. (For example, pretreatment with inducers such as DDT and phenolbarbitone increases both the amount of heme lost and the coloration whereas pretreatment with drug-metabolizing inhibitors decreases both[22]). This question was settled by the treating the post-mitochondrial supernatant of liver homogenate or the isolated microsomal fraction with AIA, which also produces the brown-green coloration, because no *de novo* synthesis can take place.[27]

The next question was whether the AIA itself is active or a metabolite. The results of drug-metabolizing stimulators or inhibitors cited above are at least consistent with the need to form a metabolite. More convincing were *in vitro* experiments which demonstrated that all the constituents for drug metabolism – NADPH, cytochrome P-450 reductase and purified lipid fraction – were required for "green-pigment" formation.[27-29] The action of cytochrome P-450 as a monoxygenase, the ability of allyl groups to from epoxides and the activity of epoxides as alkylating agents led De Matteis and Levin to suggest that the allyl-containing drugs that cause cytochrome P-450 destruction act by way of epoxide metabolites.[23,30]

Several experiments with structurally-related compounds containing propyl instead of allyl groups demonstrated the necessity of the double bond for activity. Conversion of the homoallylic amide group to a methyl ester, a methyl ketone, or a hydrocarbon terminus does not eliminate activity but saturation of the double bond does.[31] Two other types of drugs, certain dihydrocollidines (especially 3,5-diethoxy-carbonyl-1,4-dihydrocollidine, or DDC) and grisoefulvin, showed similarities with AIA and were being studied by several groups during this period. As described in the previous chapter, the first reports of the effects of DDC and griseofulvin related to hemes were drastically increased production of porphyrins (a "porphyric response"), shown to be due to blockage of chelatase activity.[32,33] It was only when differences in the behavior of these classes of porphyria-inducers became evident that their effect on cyto-chrome P-450 levels was investigated. Those collidine derivatives which cause heme destruction were found to have a common structural feature: reduction at the N and C(4) positions of the heterocyclic ring (see Figure 7.1).[34,35] Likewise, heme destruction by griseofulvin was arrested by substitution of a methoxy group by a thioether moiety.[35,36]

The product of heme alkylation in the case of AIA was proposed to be an N-alkylated derivative of protoporphyrin. The evidence for this proposal included visible-uv spectral similarities with N-methyl octaethylporphyrin and N-methyl etioporphyrin[37] and the incorporation of a ^{13}C label from the drug to the modified porphyrin product.[38] The structure was established in 1984 by Ortiz de Montellano and coworkers.[39] From the evidence that the mass spectrum gave a molecular weight equal to protoporphyrin IX (as the di-methyl ester after the workup) + AIA + an oxygen atom − ammonia,[40,41] and the fact that the product has the visible-uv spectrum characteristic of an N-alkylprotoporphyrin, they had postulated one of the two lactone structures shown in Figure 7.2.

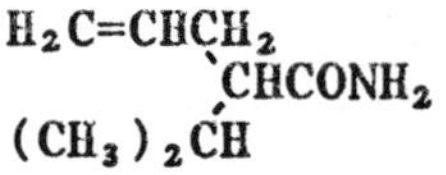
H2C=CHCH2
CHCONH2
(CH3)2CH

2-Allyl-2-isopropylacetamide (AIA)

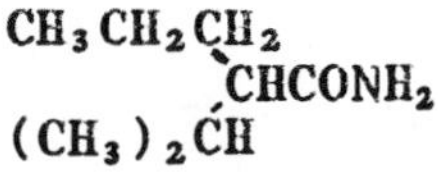
CH3CH2CH2
CHCONH2
(CH3)2CH

2-Propyl-2-isopropylacetamide (PIA)

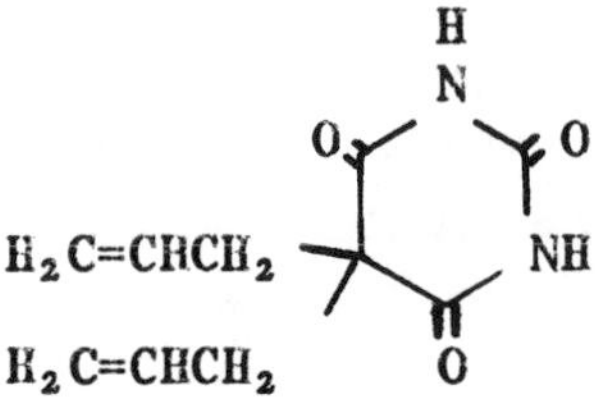
H
N
O
O
H2C=CHCH2
NH
H2C=CHCH2
O

5,5-Diallylbarbituric acid

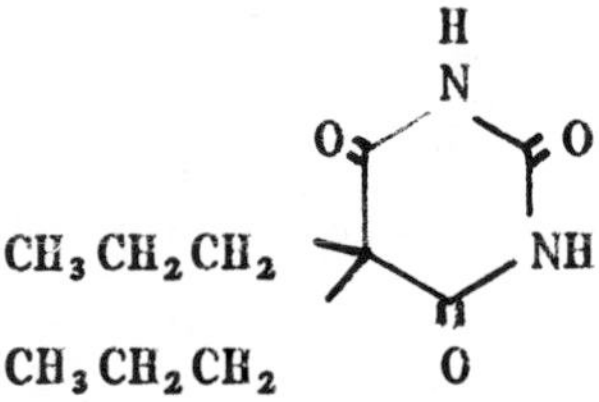
H
N
O
O
CH3CH2CH2
NH
CH3CH2CH2
O

5,5-Dipropylbarbituric acid

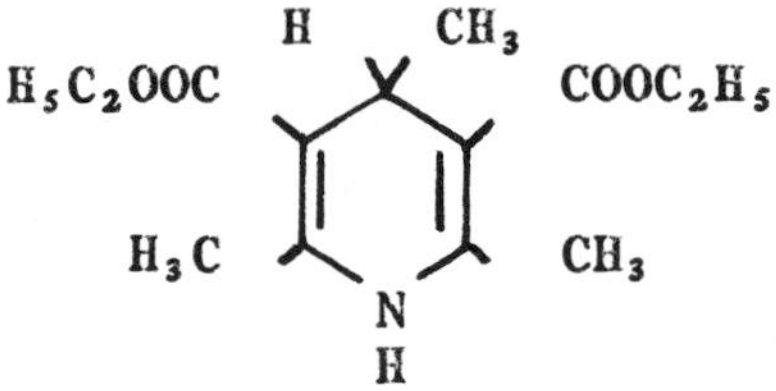
H
CH3
H5C2OOC
COOC2H5
H3C
CH3
N
H

3,5-Diethoxycarbonyl-1,4-dihydrocollidine (DDC)

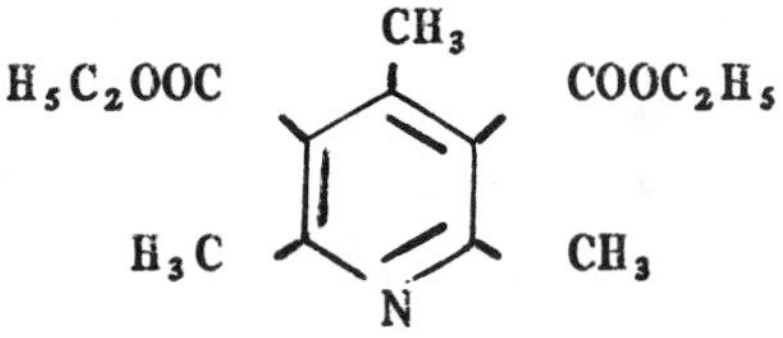
CH3
H5C2OOC
COOC2H5
H3C
CH3
N

3,5-diethoxycarbonylcollidine (DC)

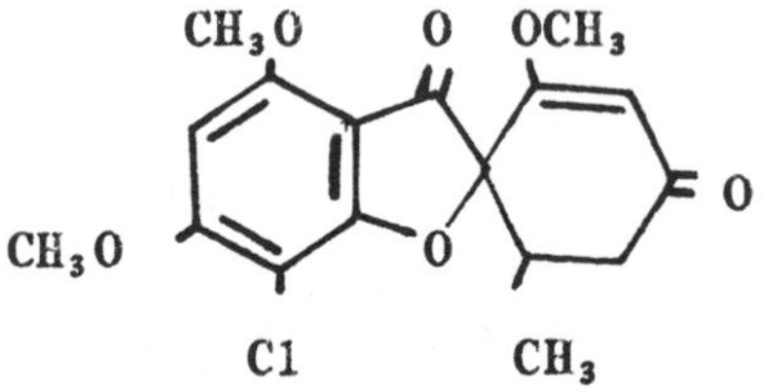
CH3O
O
OCH3
O
CH3O
O
Cl
CH3

Griseofulvin

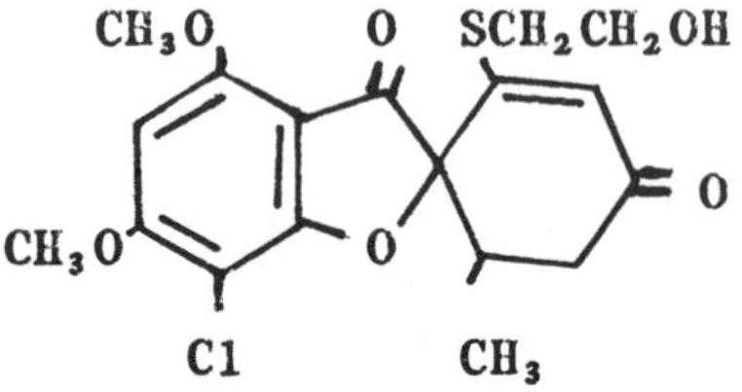
CH3O
O
SCH2CH2OH
O
CH3O
O
Cl
CH3

2-hydroxyethylthiogriseofulvin

Figure 7.1. Four classes of compounds that induce hepatic porphyria in rodents and cause destruction of hepatic cytochrome P-450 enzymes (left). The structurally-related compounds shown to the right do not have these effects.

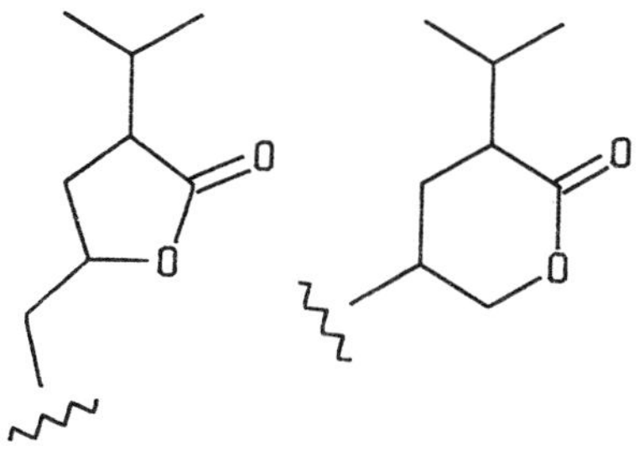

Figure 7.2. Two possible N-substitutents of protoporphyrin IX resulting from alkylation of cytochrome P-450 **in vivo** by 2-isopropyl-4-pentenamide (AIA).

The authors had noticed that the green pigment resulting from treatment of animals with 2,2-diethyl-4-pentenamide (novonal, a sedative-hypnotic available for several years as an over-the-counter drug and now available by prescription) has essentially the same visible absorption spectrum as the product from AIA treatment and that its mass spectrum shows the same relationship (mass = PP IX DME + novonal + oxygen – ammonia). The adduct from novonal is expected, moreover, to give a simpler NMR spectrum because the adduct should not possess chiral carbons like those present in the two possible products from AIA. Selective decoupling procedures and NOE results for a solution of the zinc(II) complex of the product(s) isolated from the livers of rats pretreated with phenobarbitol and then treated with novonal established the structure of the N-substituent to be that shown in Figure 7.3. From further NOE experiments, which probed the *meso* protons of the porphyrin ring, they found that there were two isomers. These isomers are the enantiomers of the chiral carbon formed upon introduction of the oxygen atom and are not due to substitution on different pyrroleninic rings. The data are consistent with substitution only on ring D, which contains the 7-propionate, 8-methyl ring substituents. From the data of the novonal derivative, they were then able to interpret the NMR spectra of the AIA derivative in terms of structure shown in Figure 7.3.

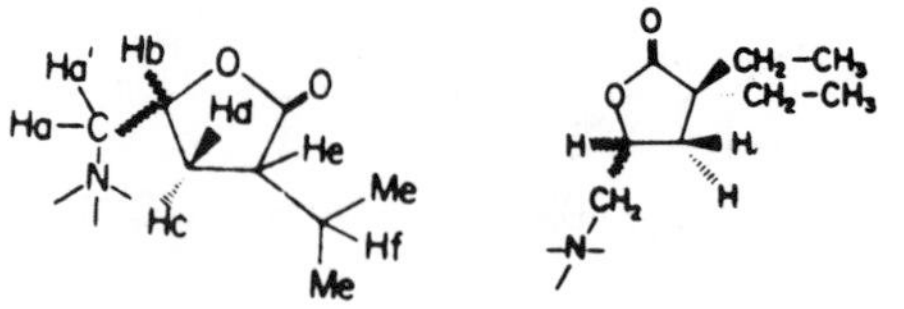

Figure 7.3. The structures of the N-substituents of protoporphyrin IX resulting from treatment of rats with novonal (left) or 2-isopropyl-4-pentenamide (AIA) (right).

on ring D, which contains the 7-propionate, 8-methyl ring substituents. From the data of the novonal derivative, they were then able to interpret the NMR spectra of the AIA derivative in terms of structure shown in Figure 7.3.

Thus, the mechanism previously proposed by Ortiz de Montellano and coworkers,[42] by which an iron-coordinated alcohol with an adjacent carbonium ion could either form the N-alkylated porphyrin by attack on the pyrroleninic nitrogen atom or the epoxide product by closure of the ring, is supported by this structural result.

The inactivation of hepatic cytochrome P-450 has been reported to be specific to isozymes induced by phenolbarbitol.[43,44] It had been suggested that perhaps only one phenolbarbitol-induced isozyme was susceptible to AIA deactivation.[44] However, later evidence has shown that several isozymes can be deactivated.[45] Hemin incubation of these isozymes (after they have lost heme because of N-alkylation) restores some activity but the degree of restoration is different for N-demethylase activity toward benzphetamine and ethylmorphine. This variation in the degree of restoration indicates that different isozymes are affected to different extents by AIA.[45]

Table 7.1. Olefins that Alkylate Cytochrome P-450 Heme.

Olefin	N-substituent	Ref.

Simple Olefins

Olefin	N-substituent	Ref.
ethylene	CH_2CH_2OH[a]	46
propene	$CH_2C(OH)HCH_3$[a]	46
1-heptene	$CH_2C(OH)H(CH_2)_4CH_3$[a]	31
1-octene	$CH_2C(OH)H(CH_2)_5CH_3$[a]	47
4-ethyl-1-hexene	$CH_2C(OH)(H)C(C_2H_5)CH_2CH_3$[b]	48
3-methyl-1-octene	$CH_2C(OH)(H)C(CH_3)H(CH_2)_4CH_3$[b]	31
vinyl fluoride	CH_2COH[a]	49
vinyl chloride[c]	CH_2COH[b]	49
vinyl bromide	CH_2COH[b]	49
ethyl vinyl ether	CH_2COH[b]	49
2,2,2-trifluoroethyl vinyl ether (fluroxene)	CH_2COH[a]	49
vinylidene fluoride	CH_2COH[a]	49
vinyl cyclooctane[d]		50

AIA Derivatives and Allylic Barbiturates

Olefin	N-substituent	Ref.
2-isopropyl-4-pentenamide(AIA)	$5-CH_2-3-isopropyl-furane-2-one$[a]	39
2,2-diethyl-4-pentenamide	$5-CH_2-3,3-diethyl-furane-2-one$[a]	39
2-phenyl-4-pentenamide	$5-CH_2-3-phenyl-furane-2-one$[b]	51
3-isopropyl-5-hexene-2-one		31
methyl 2-isopropyl-4-pentenoate		31
methyl 2-isopropyl-4-pentenylether		31
secobarbital		37,52
(5-allyl-5-(1-methylbutyl)barbituric acid		
Allobarbital[c]		24,51,53
Aprobarbital[c]		24
Sedormid		26
triallyl cyanurate[c]		53

[a] Structure confirmed by NMR spectroscopy.

[b] Structure inferred from similar substrates, visible-uv spectra have been obtained.

[c] Heme alkylation has been inferred, but no alkylated porphyrin has been isolated or identified spectroscopically.

[d] Several vinyl cycloalkanes inactivate cyt. P-450, but N-alkylporphyrins have not been characterized (Ref. 54-57).

Other olefins. In addition to the allylbarbiturates, a number of simple terminal alkenes, including ethylene, produce N-alkylprotoporphyrins by the destruction of cytochrome P-450 *in vivo* (Table 7.1). Following their report in 1980 that the visible spectrum of the product with ethylene is characteristic of an N-alkylprotoporphyrin,[31] Ortiz de Montellano and coworkers demonstrated in 1981 by NMR spectroscopy that the N-substituent is $-CH_2CH_2OH$.[46] They used both NMR spectroscopy and circular dichroism to show that the alkylation is regioselective (alkylation of a ring with a propionate side chain, ring D, being highly preferred) and stereoselective.[47]

The strikingly high specificity for one of the faces of the porphyrin was used by Montellano and coworkers to deduce the topology of the active site of cytochrome P-450. The orientation could be ascertained from previous work in which they had shown that the products from treatment of rats with 3,5-bis(carbethoxy)-2,4-dimethyl-4-ethyl-1,4-dihydropyridine (DDEP) and from the treatment of hemoglobin with ethylhydrazine were both N-ethylprotoporphyrin with the same circular dichroism spectrum (Figure 7.4) and, hence, stereochemistry.[58-60] The stereochemistry of hemoglobin is well known and the face from which oxygen enters is the same face at which hydrazines can react with the iron atom.[60] From the results of the study of the stereochemistry of the 2-hydroxyethyl adduct, Ortiz de Montellano and coworkers proposed that the pocket for the cytochrome P-450 enzyme which reacts with ethylene lies over the half of the porphyrin ring containing rings A, D and C. The structure of P-450$_{cam}$ with camphor at the active site, determined subsequently by x-ray crystallography, is consistent with this structure.[11] Since a number of cytochrome P-450 isozymes with different substrate specificities exist, the stereochemistry of N-substituted products becomes a powerful tool for establishing topological similarities or differences among them.

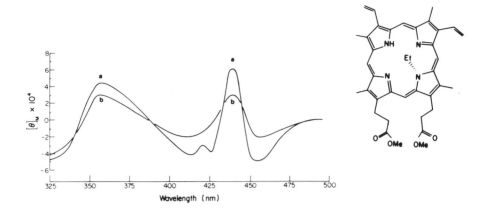

Figure 7.4. Circular dichroism spectra of a) the zinc-complexed dimethyl ester of N-(2-hydroxylethyl)protoporphyrin IX from ethylene-treated rats and b) the ring D isomer of the chlorozinc complex of N-ethylprotoporphyrin IX dimethyl ester from DDEP-treated rats[47].

The identification of the 2-hydroxyethyl group as the N-substituent produced from ethylene was followed by the investigation of a series of vinylidene derivatives to further elucidate the mechanism of N-alkylation. Ortiz de Montellano and coworkers subjected rats to treatment with vinyl fluoride, vinyl bromide and acetylene by inhalation and fluroxene and ethyl vinyl ether by interperitoneal injection; they extracted green pigments from their livers 4 hours later.[49] In the cases of vinyl fluoride, vinyl bromide, fluroxene and acetylene were all definitively shown to produce N-(2-oxoethyl)protoporphyrin IX. The fact that a common product is obtained was used to deduce the mechanistic scheme shown below (Scheme 7.5), in which a radical intermediate can provide alternate pathways, and to reject a cationic intermediate.[49] Further support for this mechanism has been obtained in model studies of reaction kinetics[61,62] and of isolated metallocyclic intermediates[17,18,62] to be discussed subsequently in this chapter.

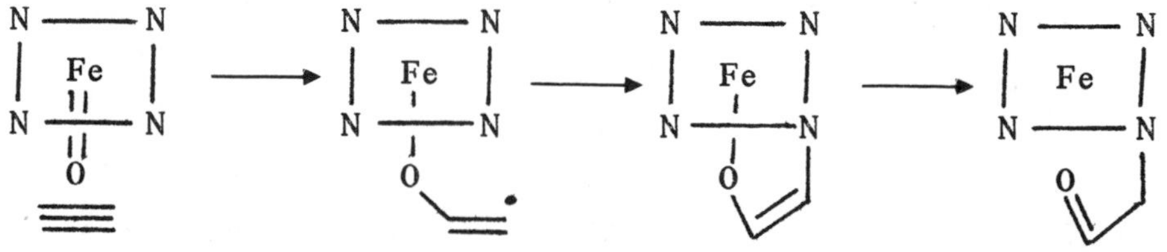

Scheme 7.5

Further development of the mechanism of heme alkylation
by olefins involved the stereochemical analysis, via stereo-
specific deuterium labelling and NMR spectroscopy, of the
epoxide and N-alkylated protoporphyrin products of the reaction of 1-octene with cytochrome P-450 in hepatic microsomes
from rat.[16,47]   The N-alkylprotoporphyrin product was
determined to be N-(2-hydroxyoctyl)protoporphyrin IX bearing
the alkyl group on the D ring, with heme alkylation occurring only on the *re* face of the olefin. (Determination of
the stereochemistry of the N-substitutent, retention of
configuration, was made by dissociating the substituent with
copper(II) ion, with the 2-hydroxy group serving as an
internal nucleophile.) The epoxide product, 1,2-oxidooc-
tane, shows only a slight excess of attack on the *re* face
(given the S enantiomer) vs. attack on the *si* face (which
gives the R enantiomer). In both cases, the oxygen atom was
shown by use of ^{18}O to arise from catalytic introduction by
the enzyme via *cis* addition. As noted previously, attack of
oxygen to form the N-alkylated product occurs at the end of
substituted olefins predicted the least reactive in corresponding epoxide metabolites.[49] In addition, epoxides of the
destructive olefins are not themselves capable of forming N-
alkyl derivatives.[42]   The stereospecificity observed leads
to two possible mechanisms: two independent paths or diver-
gence after a common acyclic intermediate.[16] In either
case, the only orientation that can lead to N-alkylation is
that in which the terminal carbon lies over the D ring and
the *re* face of the 1-octene double bond is oriented over the

reactive oxygen atom (presumably bound directly to the iron atom) of the heme. This stereoselective interaction is illustrated in Scheme 7.6.[16]

Scheme 7.6

A number of terminal olefins do not result in loss of cytochrome P-450 when incubated with microsomes of phenol-barbitol-induced rats, including styrene, 2-methyl-1-heptene and 3,3-dimethyl-1-hexene[31] and oxprenolol, alprenol and carbazeprine.[64] In addition, 1,2-disubstituted alkenes such as cyclohexene, 2-nonene and 3-hexene do not appear to alkylate the heme of cytochrome P-450.[31] Ortiz de Montellano suggested that steric congestion, electron delocalization and the presence of two carbon substituents may suppress activity, but cautioned that further work would be necessary to define the structural characteristics necessary for heme alkylation by olefins.[4]

Acetylenes. Work in this area has focussed on two major types of compounds, the ethinyl steroids and other clinically useful acetylenic agents (beginning with the 17-ethinyl sterol components of birth control pills[65] and extending to other drugs ethchlorvynol and hexapropymat[66,67]) and simple acetylenes. Some of these species are given in Table 7.2.

One of the most interesting features of the mechanism of N-alkylation by acetylenes is the finding made by Ortiz de Montellano and coworkers that it is the A ring that becomes substituted in the cases of acetylene, propyne and octyne rather than the D ring found in the case of olefins (Figure

7.5).[47] From the combination of the olefin and acetylene results, they postulated the active site orientation previously cited, with the major pocket over ring D. They explained that the difference in the bond angles of the sp^2 and sp^3 centers which fix the internal carbon of the π-bond over the iron atom (with the active oxygen atom) would place the terminal carbon atom of acetylenes over ring A and of olefins over ring D.[16] With the sedative-hypnotic agent ethchlorvynol (1-chloro-3-ethyl-1-penten-4-yn-3-ol), however, the four rings are substituted in comparable amounts.[67] The intermediate in this case may be less stereochemically restricted.

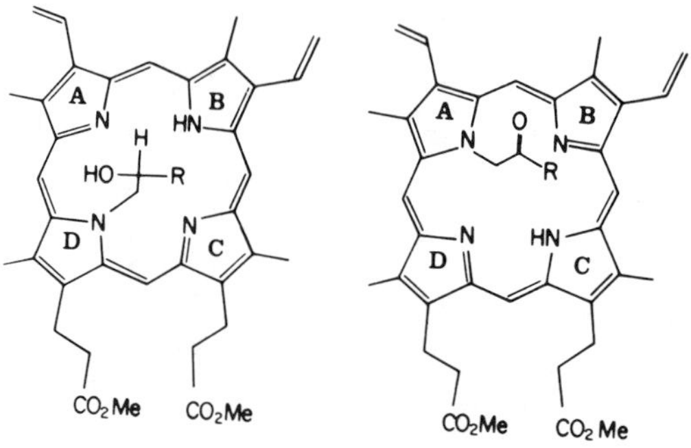

Figure 7.5. The predominant isomers formed from the interaction of olefins (left) and acetylenes (right)[47] with cytochrome P-450 in phenobarbitol-pretreated rats.

Ortiz de Montellano and coworkers have studied the relationship between acetylenes oxidation and heme N-alkylation of cytochrome P-450 to better define of cytochrome P-450 oxidation mechanisms. Using ^2H and ^{13}C labelling, they demonstrated that the enzymatic oxidation of acetylenes, like olefins, occurs by the reaction of the active oxygen bound to the highly oxidized iron atom of cytochrome P-450 with the π bond and not by insertion of the oxygen into the acetylenic carbon-hydrogen bond.[68-70] The similarity of this mechanism with that for olefins is consistent with the similar types of N-alkylprotoporphyrin products formed in the two cases.[49,71-73]

Table 7.2. Acetylenes that Alkylate Cytochrome P-450 Heme.

Acetylene	N-substituent	Ref.
acetylene	CH_2COH[a]	48,49,74
propyne	CH_2COH[a]	71
1-heptyne[b]		75
1-octyne	$CH_2C(O)(CH_2)_5CH_3$[a]	47,76
1-decyne		75
1-tridecyne[b]		75
biphenylacetylene[b]		68
cyclohexylacetylene[b]		73
1-ethinylcyclohexanol	$CH_2C(O)(C_6H_{11}-2-OH)$[c]	72,73
1-ethinylcyclopentanol	$CH_2C(O)(C_5H_{10}-2-OH)$[c]	73
(1-methoxycyclohexyl)-acetylene[b]		73
phenylacetylene		73
4-phenyl-1-butyne	$CH_2C(O)(CH_2)_2C_6H_5$[c]	73
3-(2,4-dichlorophenoxy)-1-propyne	$CH_2C(O)CH_2O(2,4-ClC_6H_4)$[c]	73
3-(4-nitrophenoxy)-1-propyne		73
3-phenoxy-1-propyne	$CH_2C(O)CH_2OC_6H_5$[c]	73
3-phenyl-1-propyne	$CH_2C(O)CH_2C_6H_5$[c]	73
2-(2-propynyl)-4-pentynamide		74
danazol		74
ethchlorvynol	$CH_2C(O)C(OH)(C_2H_5)CH=CHCl$	67,68
ethinylestradiol		65
hexapropymat[b]		64
3-methyl-1-pentyn-3-ol[b]		73
norethindrone	17α-hydrated norethindrone	15b,65
norethisterone		72
norgestrel[b]		72

[a] Structure established by NMR spectroscopy.

[b] N-Alkylporphyrin products not definitively characterized, but likely.

[c] Composition established by mass spectrometry.

They also found a large isotope kinetic effect for the
oxidation of arylacetylenes, but none for inactivation of
cytochrome P-450. (The partition ratio for oxidation and
inactivation changes from 26 to 15 when the acetylenic
hydrogen of phenylacetylene is deuterated.[70]). Thus the
path by which N-alkylation occurs for acetylenes diverges
from the path to oxidized products before the shift of the
acetylenic proton (see Scheme 7.4). Mechanisms consistent
with current evidence are shown in Scheme 7.7.[70]

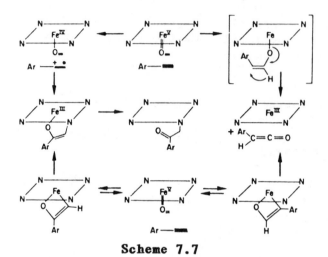

Scheme 7.7

The variety of terminal acetylenes that deactivate cytochrome P-450 is large and the reactivity of the triple bond
is apparently less easily suppressed than that of the double
bond. (Phenylacetylene[73] and biphenylacetylene[68] are reactive, but styrene is not.[31]) However, other terminal acetylenes, such as pargyline,[64] and the sedative-hypnotic ethinamate[75] are unreactive even though they are structurally
similar to active compounds (Figure 7.6). Thus, there is no
clear structure-activity relationship at this point.[4] There
are several features of heme alkylation that may make it
difficult to arrive at a simple set of structural requirements. Structural requirements for N-alkylprotoporphyrin
formation appear to be strict: the double bond must be
poised directly over the active iron-bound oxygen atom,
there are possible differences in the structure of the

pocket for P-450 isozymes and, in addition, the divergent kinetic paths for metabolite formation and N-alkylation often have rate constants of comparable magnitude. Thus, rather subtle differences are important. Mansuy has found, for example, that the cytochrome P-450 oxidation of olefins catalyzed by different reagents and oxidation catalyzed by non-enzymatic iron porphyrins can lead to aldehydes as well as epoxides and N-alkylated porphyrins and that the ratio of products depends strongly on the environment of the heme iron.[77]

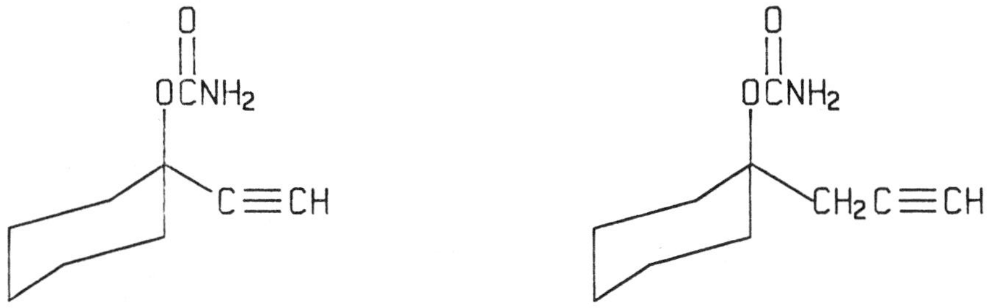

Figure 7.6. Two structurally related drugs which differ in activity toward cytochrome P-450 in rats: hexapropymat (active, left) and ethinamate (inactive, right).

White and coworkers have found that the N-substituted porphyrin produced from the contraceptive norethindrone can be converted to other N-substituted porphyrins *in vivo.*[15a] In their study they found that one green pigment (GP 1 in Figure 7.7) was produced when norethindrone was incubated with rat liver microsomes in the presence of a NADPH-generating system but that administration of the steroid to male rats gave rise to three additional N-substituted porphyrins (GP 2, 3 and 4). They also found that a cytoslic protein isolated from rat liver would cause the formation of these additional N-substituted porphyrins if it were added to the microsomal system. Antibodies raised to this protein prevented the conversion of GP1 to the other green pigments. This protein was purified, its molecular weight determined by SDS–PAGE electrophoresis and gel filtration ($M_r = 37,000$) and characterized by polyacrylamide gel isoelectric focusing (pI = 5.9). The purified protein catalyzed the reduction of Δ^4-ring-reduced norethindrone. All of these properties match those of 3α-hydroxysteroid dehydrogenase (EC 1.1.1.50).

The authors concluded that one N-substituted porphyrin with a steroid N-substituent is formed initially and that the steroid dehydrogenase enzyme is able to catalyze the hydration of the initial product despite the presence of the appended porphyrin moiety. The mass spectrographic data of the isolated green pigments[15a] is consistent with Figure 7.7.

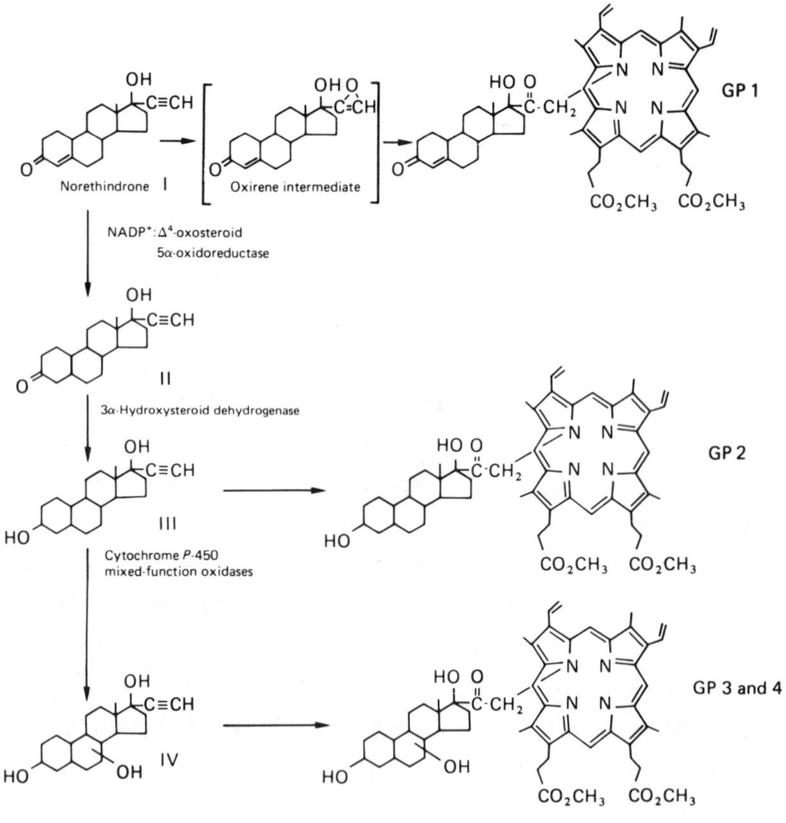

Figure 7.7. A scheme for the formation of several "green pigments" (GP 1-4) norethindrone-treated rats.[15a]

The relative reactivity of acetylenes and olefins contrasts with the behavior of allenes. Ortiz de Montellano and Kunze incubated several allenes (17-α-propadienyl-19-nortestosterone, 1-propadienylcyclohexanol and 1,1-dimethyl-allene, Figure 7.8) with microsomes from phenobarbitol-treated rats. Although the cytochrome P-450 content diminished rapidly upon exposure to these substances, there was no detectable formation of N-alkylated products.[78] The destruction of cytochrome P-450 by these allenes was shown to require catalytic turnover of the enzyme (NADPH was necessary) and it was selective for particular forms of the enzyme. (The P-450 from 3-methylcholanthrene-induced rats was not destroyed whereas the P-450 from phenobarbitol-induced rats was.) Like the acetylenes and olefins, the loss of cytochrome P-450 was matched by an equimolar loss of microsomal heme even though green pigments were not found. The authors suggested that the loss of cytochrome P-450 due to these compounds may be a result of attack of the heme by a highly unstable intermediate such as an *exo*-methylene epoxide (for example, as in the "bleaching" of heme by cumene hydroperoxide, hydrogen peroxide[79] and spironolactone[80] or that the apoprotein could be the target of a highly reactive transient intermediate[78] (as, for example, proposed by Mansuy in the case of destruction of cytochrome P-450 by chlorohydrocarbons[81]). It is quite possible that N-alkylated products are produced but that they are relatively unstable. If they have any importance in the destruction of the P-450 enzymes, of course, they would have to persist long enough to be extruded from the enzyme or they would have to be intermediates by which the allenes then attack the apoprotein. It would be interesting to obtain data on model systems to investigate the reactivity of such species and also to determine the chemical structure of the lost heme.

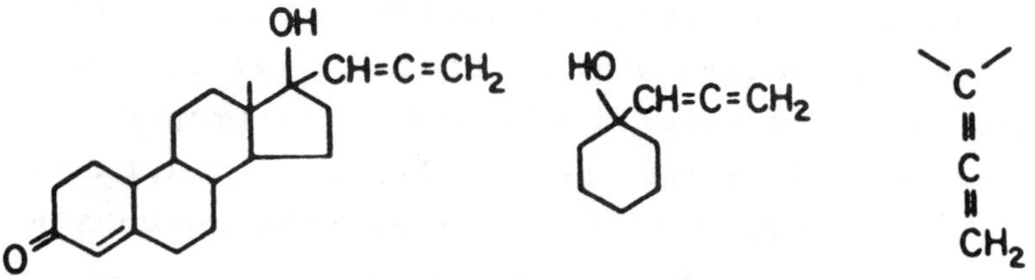

17-α-propadienyl-   1-propadienyl-   1,1-dimethylallene
19-nortestosterone   cyclohexanol

Figure 7.8. Three allenes which destroy cytochrome P-450 but do not produce large amounts of N-alkylated products.[78]

4-Alkyl-1,4- Dihydropyridines. The observation by De Matteis and coworkers that 3,5-diethoxycarbonyl-1,4-dihydrocollidine (DDC), which had been known to induce hepatic porphyria, causes a decrease in ferrochelatase activity[82] was followed by a search for the identity of the responsible agent. Much of the background work of DeMatteis and his group on the spectral comparisons of the "green pigments" produced by DDC with known N-alkylporphyrins,[83,84] the use of radiolabels to demonstrate the origin of the N-alkyl group as the C(4) substituent,[85,86] and the varied effect of dihydropyridines bearing other C(4) substituents (Table 7.3)[87-89] has been discussed in the previous chapter as it related to studies of ferrochelatase inhibition. It should be noted that three types of behavior of DDC derivatives have been observed: 1) N-alkylporphyrin production and ferrochelatase inhibition, 2) N-alkylporphyrin formation but no ferrochelatase inhibition, and 3) neither phenomenon, depending on the 4-substitutent.[90] Ortiz de Montellano and coworkers deduced from NMR data that DDC administered to phenolbarbitol-induced rats produces N-methylprotoporphyrin IX that is methylated principally on the A ring.[59,91]

The mechanism for transfer of the 4-methyl (or other C(4) substitutents for active analogs) involves oxidation of the substrate by the enzyme; therefore, non-reduced pyridines are inactive.[85,87,92] Marks and coworkers have shown that the 2-ethoxycarbonyl functionality is necessary to preserve the reduced dihydropyridine structure in a living system.[147] They found that when 3-ethoxycarbonyl-1,4-dihydro-2,4-dimethylpyridine instead of DDC is administered to chick embryo liver cells in culture, there was much less ferrochelatase inhibition and the types of excess porphyrin that were formed differed. They noted that it is known that 3-ethoxycarbonyl substituted dihydropyridines are rapidly transformed in aqueous solution to nondihydropyridine products. As discussed in the previous chapter, the interference with heme biosynthesis by way of ferrochelatase inhibition is directly linked to heme alkylation. Cytochrome P-450 is deactivated with concomitant N-alkylporphyrin formation when there are ethyl, propyl, isopropyl, isobutyl or benzyl moieties in place of the 4-methyl group, but N-alkyl derivatives have not been found for the latter two analogs.[88,93] If iron complexes with organic axial ligands serve as intermediates, they may not be sufficiently stable to allow N-alkylation in these cases or it may be that the activated complex for N-alkylation would be difficult to achieve since the isopropyl N-alkyl complex would be relatively bulky and the N-benzyl complex relatively unstable with respect to dealkyaltion. Ortiz de Montellano and coworkers demonstrated by EPR spin-trapping that oxidation of the 4-ethyl analog by microsomal cytochrome P-450 releases the 4-ethyl group as a free radical;[93] they concluded , therefore, that production of N-ethylprotoporphyrin suggests that a catalytically-released radical reacts with the prosthetic heme group and inactivates the enzyme.[5] This is consistent with results for N-substituted porphyrin formation by hydrazines, discussed in detail in the following chapter.

Table 7.3 Heteroatomic Substrates that Deactivate Cytochrome P-450 by N-Substitution of the Prosthetic Heme.

Compound	N-Substituent	Ref.
4-alkyl-1,4-dihydropyridines		
3,5-bis-carbethoxy-2,4,6-trimethyl-1,4-dihydroxypyridine (DDC)	$-CH_3$[a]	59
3,5-bis-carbethoxy-4-ethyl-2,6-dimethyl-dihydroxypyridine (DDEP)	$-CH_2CH_3$[a]	94
3,5-bis-carbethoxy-4-isopropyl-4,6-dimethyl-dihydroxypyridine	$-CH_2CH_3$[b]	88,93
Hydrazines		
phenylhydrazine	$-C_6H_5$[b]	95
2-phenylethylhydrazine (phenelzine)	$-CH_2CH_2C_6H_5$[a]	96
Nitrosamines		
dimethylnitrosamine		97
dipropylnitrosamine		97
Aryne and Cyclobutene Precursors		
1-aminobenzotriazole	$1,2-C_6H_4$[a,c]	98
N-acetylmethyl, N-acetylmethyl, N-dimethyl, and N-methyl-1-aminobenzotriazoles	$1,2-C_6H_4$[b]	98
N,N'-bis-carbethoxy-2,3-diaza bicyclo[2,2,0]hex-5-ene	$-C_4H_5$[a,c]	99

a   Structure established by NMR spectroscopy.
b   N-Alkylporphyrin products not definitively characterized.
c   The product is N,N'-phenylprotoporphyrin IX.
d   The product is N-(2-cyclobutenyl)protoporphyrin IX.

Hydrazines. The important consequence of the destruction of cytochrome P-450 by a hydrazine has been demonstrated by the drug meperidine, which becomes highly toxic when administered with 2-phenylethylhydrazine (phenelzine).[100,101] The phenelzine irreversibly inactivates cytochrome P-450,[102] forming N-(2-phenylethyl)protoporphyrin IX.[96] Ortiz de Montellano and coworkers have used spin trapping experiments to demonstrate that the mechanism for formation of the N-substituted product involves oxidative formation of phenylethyl free radical, which presumably attacks the heme prosthetic group.[96] In view of the reactions of phenylhydrazine with other hemoproteins such as hemoglobin, myoglobin and catalase[103-105] the reaction of phenylhydrazine with cytochrome P-450 probably produces N-phenylprotoporphyrin IX. The presence of a transient intermediate in the reaction which absorbs at 449 nm has been reported.[95] A number of other hydrazines which do not seem to produce N-substituted products have also been shown to produce intermediates that absorb at 449 nm (including isoniazid and hydralazine[106-109] and 1,1-disubstituted hydrazines.[110] The mechanism of these reactions may involve diazene complexes, since Mansuy and coworkers have found that direct reactions between diazenes and heme *in vitro* produce complexes with an absorbance at 447 nm.[111] Extensive studies have been performed on the reactions of hydrazines with a number of hemeproteins and many interesting model studies have also been reported (see Chapter 8).

Nitrosamines. The N-nitrosamines are tissue-specific carcinogens which undergo oxidative dealkylation by tissue-specific microsomal oxygenases. In this process, active alkyl cations are formed which bind to DNA or other nucleophilic molecules. It has been proposed that the dealkylations result from hydroxylations of the α-carbons by mixed function oxygenases and there is evidence that cytochrome P-450 isozymes differ in their activity toward different dialkylnitrosamines, which provides a rational explanation for the tissue-specificity shown for different alkyl substi-

tuents.[112] The general properties of the dialkylnitrosamines are related to the formation of N-alkyl protoporphyrin IX *in vivo*. It has been known for a considerable time
that nitrosamines reduce the concentration of cytochrome P-
450 enzymes *in vivo*, thereby inhibiting the metabolism of
drugs.[113] Recently, White and coworkers have shown that N-
substituted products are formed *in vivo* from a number of
nitrosamines, including dimethyl-, diethyl- and dipropylnit-
rosamine.[96] In the case of diethylnitrosamine, the product
was identified as N-(2-hydroxyethyl)protoporphyrin IX, the
same product as produced from ethylene (discussed above).
As in the case of ethylene, no N-ethylprotoporphyrin IX was
isolated, indicating that the mechanism of formation of the
product proceeds via an oxidized intermediate. The mechanism for formation of N-(2-hydroxyethyl)protoporphyrin IX
from ethylene that has been proposed by Ortiz de Montellano
(Schemes 7.2 - 7.4) involves an oxidized cationic intermediate. The fact that alkyl alcohols required for the formation of such intermediates are known metabolites of dialkyl-
amines makes a similar mechanism plausible for nitrosamine
deactivation of cytochrome P-450.[114]

Aryne and Cyclobutadiene Precursors. Chemical oxidation
of 1-aminobenzotriazole leads to benzyne formation and its
interaction with cytochrome P-450 in hepatic microsomes from
phenolbarbitol induced rats leads to a product in which a
benzene moiety spans two of the nitrogen atoms of protoporphyrin IX (Figure 7.9).[69,98,114]

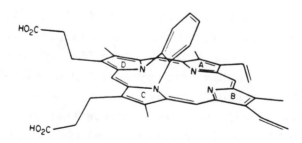

Figure 7.9 N,N'-(1,2-phenylene)protoporphyrin IX.[98]

The behavior of other unsaturated substrates such as alkenes and alkynes, which produce N-substituted protopor-phyrin IX via oxidized intermediates, makes plausible the mechanism proposed by Ortiz de Montellano and coworkers which invokes a benzyne intermediate (Scheme 7.8).[98] They performed experiments with a number of substituted benzo-triazoles, as shown in Table 7.3, and found that the pro-ducts all exhibited the same chromatographic and spectrosco-pic behavior; they are, therefore, all likely to be N,N'-(1,2-phenyl)protoporphyrin IX. The insecticide synergist 1,2,3-benzothiadiazole may not deactivate cytochrome P-450 in the same manner since no N-alkylprotoporphyrin product has been identified.[116]

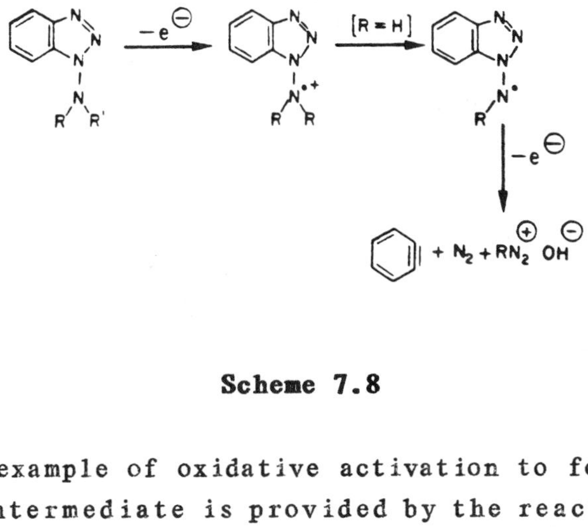

Scheme 7.8

Another example of oxidative activation to form a highly reactive intermediate is provided by the reaction of 2,3-bis-2,3-diazabicyclo[2.2.0]hex-5-ene, which produces cyclo-butadiene by chemical oxidation, with cytochrome P-450 in hepatic microsomes from phenolbarbitol-induced rats. This reaction produces N-(2-cyclobutenyl)protoporphyrin IX, identified unambiguously by NMR spectroscopy to be bound to the D ring of the porphyrin.[98,117] Of the several possible routes shown in Scheme 7.9, Stearns and Ortiz de Montellano have argued convincingly for the primary role of the radical cation intermediate based on the known chemistry of the

substrate and the relative stabilities of electrochemically
produced cation radicals of analogous compounds.[98] They
have pointed out that the catalytic release of cyclo-
butadiene in this case and of benzyne in the case of the
benzotriazoles makes it possible to contemplate the use of
such stereoelectrically activated intermediates to design
mechanism-based enzyme inhibitors.[98]

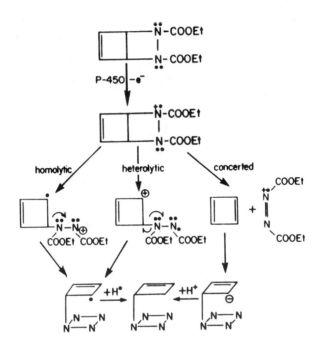

Scheme 7.9

Sydnones. This class of diazaheterocyclic compound,
which has analgesic, antimalarial, antifungal and antiin-
flammatory activity,[118,119] has not been clinically useful
because animal tests have resulted in massive accumulations
of liver pigments that have been tentatively been identified
as protoporphyrin IX.[120-122] In 1986, Ortiz de Montellano
and Grab reported that the administration of a sydnone, 3-
{[2-(2,4,5-trimethylphenyl)thio]ethyl}-4-methylsydnone
(TTMS), to phenolbarbitol-pretreated rats produces N-vinylprotoporphyrin IX.[123] They definitively identified all four
isomers of the compound (Figure 7.10) isolated from the

livers of rats exposed to TTMS, which were converted to dimethyl esters and then purified by HPLC, by using NMR spectroscopy. The schemes Ortiz de Montellano and Grab used to rationalize the formation of N-vinylprotoporphyrin IX from cytochrome P-450 were based upon the oxidation mechanism proposed for chemical oxidation of syndones by White and Egger[124] (Scheme 7.10) and the model studies by Johnson, Callot, and Mansuy and their coworkers which have shown that diaza compounds can produce iron–nitrogen bridged intermediates (discussed in detail in Chapter 8), as shown in Scheme 7.11.

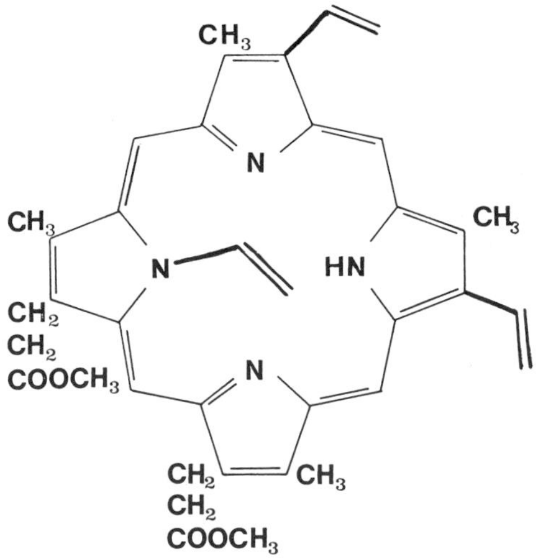

Figure 7.10. The dimethyl ester of one of the isomers of N-vinylprotoporphyrin IX, formed from a sydnone.[123]

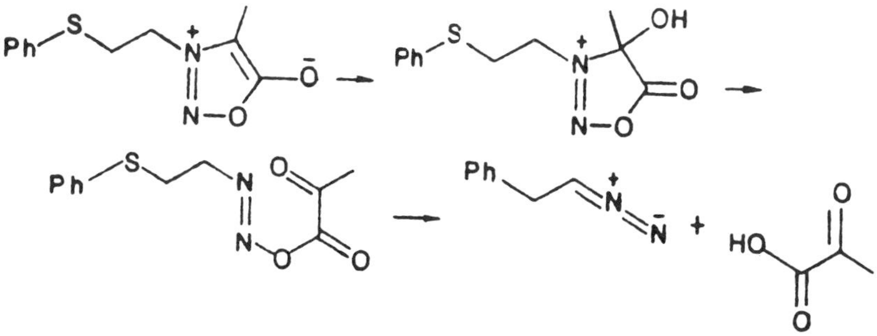

Scheme 7.10

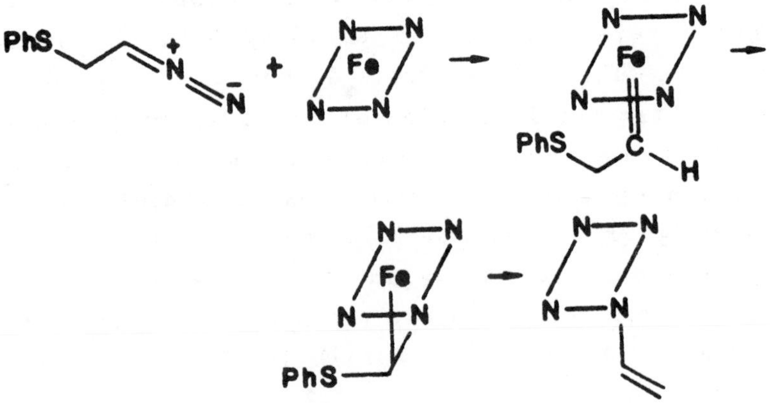

Scheme 7.11

Production of excess protoporphyrin IX upon administration of sydnones (which form of N-vinylprotoporphyrin IX) is reasonable since several groups have found that N-alkyl-protoporphyrins having a small N-substituent such as methyl or ethyl greatly inhibit ferrochelatase (Chapter 6). Protoporphyrin IX then accumulates as the organism tries in vain to manufacture heme. Since the vinyl substituent is somewhat less bulky than the ethyl group, it also would be expected to inhibit ferrochelatase. An interesting aspect of Ortiz de Montellano and Grab's study of the action of sydnones on test animals is how it ties together a diverse range of studies of N-substituted porphyrins: they have applied results from pharmacological studies of other drugs that produce N-alkylporphyrins by different mechanisms to interpret their spectroscopic data (Chapter 3), referred to the model studies of Callot and others on synthetic cobalt porphyrins (Chapter 8) to rationalize the mechanism of action and used studies of the effects of N-alkylporphyrins on ferrochelatase (Chapter 6) to explain the physiological consequences of administering the drug to test animals.

Other Substrates that Inhibit Cytochrome P–450 but May Not Produce N–Substituted Protoporphyrin IX. There are a number of substrates for cytochrome P–450 enzymes that cause loss of microsomal heme but for which no definitive evidence has reported to demonstrate alkylation of the prosthetic heme. Among these substrates that may cause alkylation or formation of relatively unstable N–substituted protoporphyrin derivatives but for which conclusive evidence is unavailable are aldehydes, cyclopropylamines, halogenated hydrocarbons, thiocarbonyls and thiophosphates.[5]

Aldehydes. Patel and coworkers have demonstrated that aromatic aldehydes cause an equimolar loss of microsomal cytochrome P–450 *in vivo,* whereas aliphatic aldehydes only cause loss *in vitro.*[125–127] As in the cases of substrates that produce N–substituted protoporphyrin via enzyme oxidized intermediates, the loss of cytochrome P–450 only results under conditions that are required for catalytic turnover (that is, both NADPH and oxygen must be present). The formation of a "green pigment" has been reported upon incubation of octanal with hepatic microsomes from rat, but the visible absorption spectrum of the pigment that was isolated is not like those of that are commonly found for well–identified N–alkylprotoporphyrins.[76]

Cyclopropylamines. Cytochrome P–450 enzymes are irreversibly inhibited by benzylcyclopropylamines. The proposed mechanism involves oxidation of the cyclopropylamine to an exocyclic imine and cleavage of the cyclopropyl ring to form a carbon radical intermediate.[128,129] The carbon radical could deactivate the enzyme either by attacking the heme to form an N–alkylprotoporphyrin or by attacking the protein to form a funtionally deficient derivative. Definitive evidence for N–alkylprotoporphyrin formation has not yet been reported although formation of a "green pigment" was mentioned.[129]

Halogenated hydrocarbons. The interactions of carbon tetrachloride and the anesthetic halothane (1,1,1-trifluoro-2-bromo-2-chloroethane) with cytochrome P-450 have been extensively studied. In model studies using the iron(II) complex of tetraphenylporphyrin, Mansuy and Battioni demonstrated that halothane can form a σ-alkyl product (the axial ligand being -CHClCH₃, the only σ-alkyl complex of an iron porphyrin with a halogen substituent on the carbon atom bound to iron) under the same conditions where carbon tetrachloride forms a carbene complex.[130] In each case, a reversible intermediate which absorbs at long wavelengths (454 nm for carbon tetrachloride and 469 nm for halothane) is formed when the substrates are incubated with hepatic microsomal cytochrome P-450 but, unlike the case of the diazene intermediates in hydrazine reactions, these interme-diates are not well identified.[62,131-134]

When incubated with rat microsomes, benzyl halides (benzyl bromide and p-nitrobenzyl chloride) also form long-wavelength-absorbing intermediates (478 nm), deduced to be σ-alkyl ferric cytochrome P-450 complexes.[135] No N-alkyl-protoporphyrin IX products were isolated from these reactions, possibly because the σ-alkyl intermediates are too unstable to survive long enough for the metal-to-nitrogen migration reaction to occur.

Considerable work by Mansuy and coworkers indicates that polyhalogenated hydrocarbons can form carbene-heme com-plexes.[81] At least in the case of DDT (1,1,2-trichloro-bis(p-chlorophenyl)ethane), the carbene complex[136] can undergo a migration reaction to form an N-vinyl porphyrin (in the case of DDT and tetraphenylporphyrin, N-[2,2-bis(p-chlorophenyl)vinylidene]tetraphenylporphyrin[137,138]). Although products from the incubation of carbon tetra-chloride with hepatic microsomes have yielded radiolabelled products that could be N-substituted products,[139] no fluorescent N-alkylated products were obtained under workup conditions which normally give fluorescent products (free base N-alkylporphyrins) after allylisopropylacetamide treatment.[131] It is possible that only those carbenes that have highly stabilizing substituents undergo migration and that other carbenes, such as the dichloromethine derived from carbon tetrachloride, inactivate cytochrome P-450 by attacking the protein rather than the heme moiety. The various possibilities for enzymatic degradation by halogenated hydrocarbons have been discussed by Mansuy.[81] Lange and Mansuy have speculated that the destruction of cytochrome P-450 by halogenated hydrocarbons, as well as by the insecticide synergists of that are 1,3-benzodioxoles, may react like DDT, forming N-substituted products,[137] but the deactivation mechanism for these compounds is unclear.

Thiocarbonyls and thiophosphates. Several commercially important insecticides include C=S and P=S functionalities which very likely act by deactivating the detoxification apparatus of insects. It has been demonstrated that compounds such as diethyldithiocarbamate and parathion (diethyl-p-nitrobenzoylthiophosphinite) inactivate cytochrome P-450 and cause loss of microsomal heme under catalytic conditions.[2,10,140-142] For parathion, the loss of catalytic activity correlates with loss of the heme of the cytochrome P-450 enzyme.[10,142] The products of this reaction have not been identified. Mansuy and coworkers have isolated thiocarbene complexes of iron tetraphenylporphyrin[143] but the migration to form N-substituted products is not yet known. They have also shown that fungicides of the type RSCCl₃ form carbene complexes of the type Fe-CClSR with synthetic iron porphyrins. These are transformed to stable thiocarbonyl-iron porphyrin complexes in the presence of Lewis acids. Characterization of the products from interactions of the fungicides with cytochrome P-450 has not been as clear.

Model Studies Related To Cytochrome P-450 Reactions

Certainly a great deal of information about the mechanism of formation of N-substituted protoporphyrins has been achieved by direct study of reactions of cytochrome P-450 enzymes. The reactivity as a function of the terminal substituents of hydrocarbons and the stereoselectivity of both the oxidation process and the ring of the heme prosthetic group that is alkylated have defined requirements for proposed mechanisms. In some respects, however, direct study of an enzymatic system is inherently limited. The pocket in which the substrate resides and the structure of the protoporphyrin ring of the prosthetic group cannot be systematically altered to investigate these effects of reactivity. Also, the study of intermediates is limited by their inherent (in)stability. It is in these respects the model studies can contribute to a clearer definition of intimate mechanistic details.

Reactions in which iron porphyrins catalyze the oxidation of olefin to form epoxides, like reactions of cytochrome P-450, sometimes terminate when the catalyst is converted into an N-alkylporphyrin. Synthetic reactions also mimic the stereoselectivity of epoxidation found for enzymatic reactions. Similarities between the model and biochemical reactions and possible mechanisms for explaining the partitioning of the substrate to either N-alkylation of the heme prosthetic group (suicidal inhibition) or epoxidation have been discussed by Mansuy.[144] He and his coworkers used rather simple synthetic porphyrin complexes as catalysts (chloroiron(III) complexes of tetraphenylporphyrin and p-chlorotetraphenylporphyrin) and iodosylbenzene as oxidant (to mimic cytochrome P-450 and dioxygen, respectively). Systems of this sort had previously been studied by Groves and others.[145] Spectroscopic evidence favors a high valent intermediate formulated as an iron(IV)-oxo-porphyrin cation radical complex. Other workers had also found that the restriction to the metal atom caused by bulky groups

attached at the ortho positions of the *meso*-phenyl substituents gave products in isomeric ratios closer to the
enzymatic system.[146]   While other groups had focussed on the
sterochemistry and catalytic yields of the synthetic
systems, Mansuy and his coworkers carried the comparison
further by also determining which substrates formed N-alkylporphyrin products.   They found that there were close
parallels of the natural and synthetic systems.   Both
systems gave N-alkylporphyrin products with substrates such
as the terminal olefins with unsubstituted γ positions but
not with dialkylated γ carbons. *Cis*-2-alkenes did not give
N-alkylporphyrin products in either case. However, *trans*-2-
hexene is an exception, giving small amount of an N-alkylporphyrin product in the synthetic system but not with
cytochrome P-450. The scheme Mansuy proposed[144] is given in
Scheme 7.12 (species 2 is an intermediate of unknown
structure, presumably with the metallocyclic ring opened).

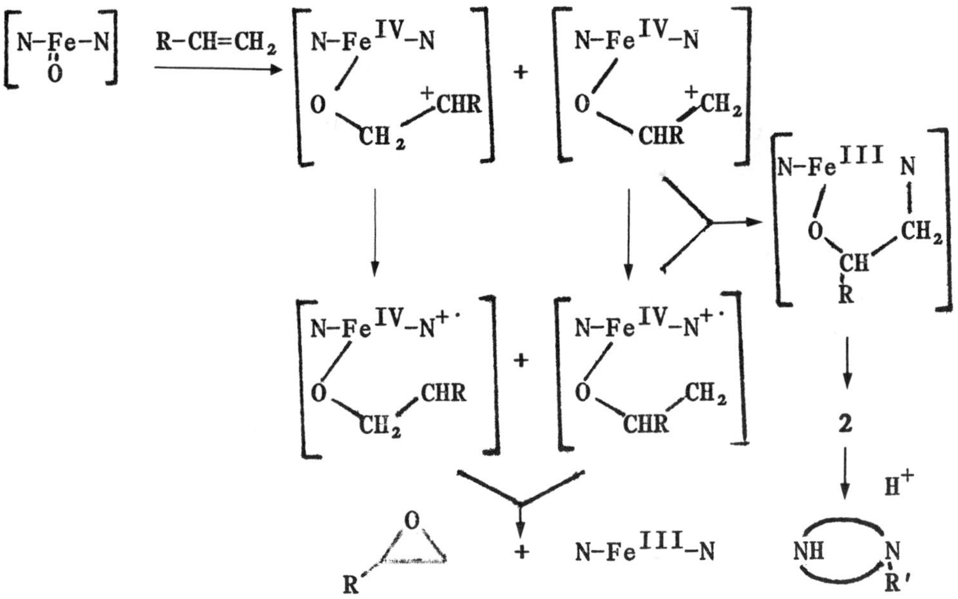

Scheme 7.12

An especially interesting aspect of the proposed
mechanism is the formation of a metallocylic intermediate of
a type unknown until recently.   Work in the laboratories of

Collman, Dolphin, Mansuy and Traylor have now provided examples of metallocyclic complexes closely resembling possible intermediates formed during the production of N-alkylprotoporphyrin IX species from cytochrome P-450 enzymes by suicide-substrates. Evidence provided by the extensive work of Ortiz de Montellano and coworkers and several other groups, discussed in the previous section of this chapter, has indicated that the moiety attacking the pyroleninic nitrogen atom of the heme prosthetic group has first been catalytically oxidized and that the oxidation is highly stereoselective. The reasonable mechanism proposed by Ortiz de Montellano and coworkers to accommodate these facts has been shown for the case of ethylene in Scheme 7.2. Until recently, there were no examples of iron porphyrin complexes that included a metallocycle linking the iron atom with a pyroleninic nitrogen atom.

In 1985, Collman and coworkers presented kinetic data for the reaction of pentafluoroiodosylbenzene with several olefins, using iron(II) tetrakis(pentafluorophenyl)porphyrin chloride as the catalyst. These data strongly suggest a metallocyclic intermediate in which an oxygen added to the β-carbon of the olefin is attached to the iron atom and the α-carbon is attached to the pyroleninic nitrogen atom.[61] Further work with this synthetic catalytic system demonstrated that it behaved similarly to the cytochrome P-450 reactions, in that the model system showed suicide porphyrin N-alkylation and stereoselective epoxidation that closely resemble those for cytochrome P-450.[16,62] In several cases, N-alkyltetrakis(pentafluorophenyl)porphyrin products were definitively characterized by NMR spectroscopy (for example, the N-substituents in the cases of 3-methyl-1-butene and methylenecyclohexane are the 2-hydroxy-3-methyl-butyl and 2-hydroxy-2-methyl-1-propyl groups, respectively.[62]) The authors studied the kinetics of these reactions and found that the rates were independent of olefin and catalyst concentrations. They also found that the partition ratios (moles of epoxide formed to moles of catalyst N-alkylated) were near those observed for cytochrome P-450 and allylisopropylacetamide (200–230) for several of the olefins studied (150 for 1-octene, 150 for 3,3-dimethyl-1-butene, 250 for vinyl cyclohexane and 800 for methylene cyclohexane).   The mechanism to account for the stereoselectivity and partitioning of the various olefins proposed by Collman and coworkers in 1986 is shown in Scheme 7.13.

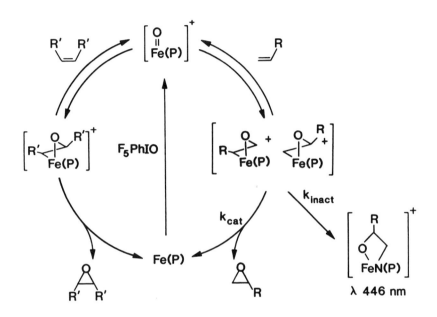

Scheme 7.13

In 1985, Dolphin and Traylor and their coworkers used a similar synthetic catalytic system to characterize a species with a visible absorption spectrum (abs. max. and log abs., 425(sh), 444 (4.70), 455 (sh), 566 (3.93), 599 (3.88), 640 (sh) 670 (sh)) and elemental analysis consistent with a metallocyclic structure.[17]   In their case, the catalyst was tetrakis(2,6-dichlorophenyl)porphinatoiron(III) and the olefin which led to the isolated species was 4,4-dimethyl-1-pentene.   The metallocycle was formed by complexing iron to

the N-alkylporphyrin product which formed as the catalyst,
N(4,4-dimethyl-2-hydroxypentyl)tetrakis(2,6-dichlorophenyl)-
porphyrin, is deactivated. In Scheme 7.14 shown below, the
metallocycle formed from the N-alkylporphyrin product (4) is
5.

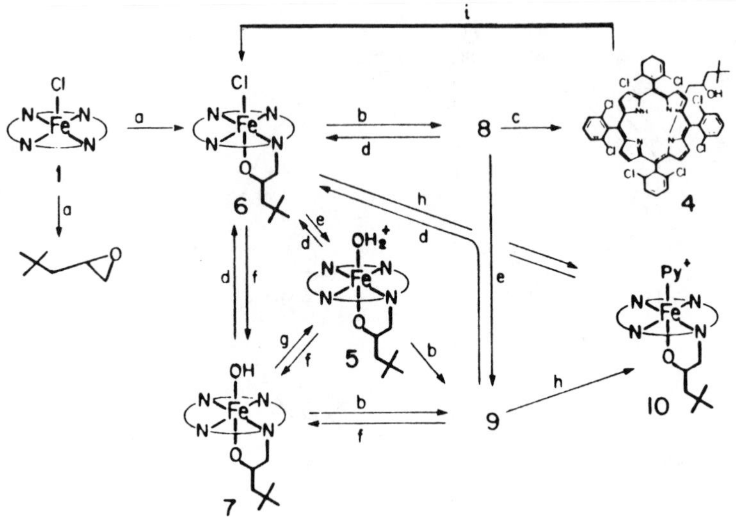

Scheme 7.14

Scheme 7.14 represents the possible interconversions of the
N-alkylporphyrins in the oxygenation of 4,4-dimethyl-1-
pentene by iodosopentafluorobenzene catalyzed by tetrakis-
(2,6-dichlorophenyl)porphyrin.[17]   The reagents are as
follows (with dichloromethane except c, f and i): **a**: 4,4-
dimethyl-1-pentene, pentafluoroiodosylbenzene, **b**: pentafluoroiodosylbenzene, **c**: concentrated HCl in acetic acid, **d**: 1%
aqueous HCl, **e**: water, **f**: aqueous sodium bicarbonate, **g**:
acetic acid, **h**: pyridine and **i**: ferrous chloride and tetrahydrofuran.

The most definitive evidence for metallocyclic iron porphyrins with N-substituents is the crystallographically-
determined structure of a bicylic complex (Figure 2.7 on
page 29).[63] Strong spectroscopic support for structures of
several monocyclic complexes, including novel six-coordinate
low spin Fe(II) complex and a Co(III) complex have been
obtained by Mansuy and coworkers.[148-150] Although these
species were formed using model complexes, they demonstrated
that such a structure is feasible for cytochrome P-450
reactions. In addition, they provide the spectroscopic
characteristics that make Dolphin and Traylor's postulated
metallocyclic structure highly plausible.[17]

An interesting feature of the model systems that Mansuy
and Traylor have studied is that they have found N-alkyl-
hemin formation for non-terminal olefins,[144,148-152] whereas
the N-alkylprotoporphyrins from cytochrome P-450 reactions
have only been obtained from terminal olefins. Mansuy and
coworkers found that a spatially hindered porphyrin (iron
tetrakis(2,6-dichlorophenyl)porphyrin) produced and N-
CH_2HOHR porphyrin attached at the *less* substituted carbon of
the alkene double bond while iron tetraphenylporphyrin
produced a product attached at the *more* substituted carbon
atom.[144,149] The products obtained with such 'secondary'
attachments are much less stable than typical N-substituted
porphyrins. The N-alkylhemins from reactions of norbornene
and cyclohexene are only observed as transients.[151,152]
Therefore, it is possible that such products are formed in
cytochrome P-450 reactions but that they either revert to
other products while still within the enzyme[151] or they
undergo reactions after dissociating from the protein but
before (or during) isolation procedures. The work of Mansuy
and Traylor's groups indicate that reactions of both model
systems and, likely cytochrome P-450 as well, are highly
complex and that N-alkylhemins may be involved for a wider
variety of substrates than previously considered.

References

1. F. De Matteis, Hemes and Hemoproteins, F. DeMatteis and W.N. Aldridge, eds., Springer-Verlag, N.Y., **44** (1978) 95-127.

2. F. De Matteis, J. Jarvisalo, and A.H. Gibbs, Molec. Pharmacol., **14** (1978) 1099-1106.

3. P.R. Ortiz de Montellano, B.A. Mico, G.S. Yost and M.A. Correia, Enzyme Activated Irreversible Inhibitors, M. Seiler, Elsevier, Amsterdam (1978) 337-352.

4. P.R. Ortiz de Montellano and M.A. Correia, Ann. Rev. Pharmacol. Toxicol., **23** (1983) 481-503.

5. P.R. Ortiz de Montellano, Ann. Rep. Med. Chem., **19** (1984) 201-211.

6. B. Testa and P. Jenner, Drug. Metab. Disp., **12** (1981) 1.

7. K.J. Netter, Pharmacol. Ther., **10** (1980) 515-535.

8. M.R. Franklin, Pharmacol Ther. A, **2** (1977) 227-245.

9. E. Hodgson and R.M. Philpot, Drug Metab. Rev., **3** (1974) 231-301.

10. J. Halpert and R.A. Neal, Drug. Metab. Rev., **12** (1981) 239-259.

11. T.L. Poulos, B.C. Finzel, I.C. Gunsalus, G.C. Wagner and J. Kraut, J. Biol. Chem., **260** (1985) 16122-16130.

12. D.K. Lavallee, J. Inorg. Biochem., **16** (1982) 135-143.

13. D. Kuila, A.B. Kopelove and D.K. Lavallee, Inorg. Chem., **24** (1985) 1443-1446.

14. F. DeMatteis, F. Gibbs and A.P. Unseld, Adv. Exp. Med. Biol., **136B** (1982) 1319-1334.

15. a) I.N.H. White, Biochem. Pharm., **31** (1982) 1337-1342.
    b) I.N.H. White, D.C. Blakey, M.L. Green, M. Jarman and H.-R. Schulten, Biochem. J., **236** (1986) 379-387.

16. P.R. Ortiz de Montellano, B.L.K. Mangold, C. Wheeler, L. Kunze and N.O. Reich, J. Bio. Chem., **258** (1983) 4208-4213.

17. T. Mashiko, D. Dolphin, T. Nakano and T.G. Traylor, J. Amer. Chem. Soc., **107** (1985) 3735-3736.

18. J.P. Battioni, I. Artaud, D. Dupre, P. LeDuc, I. Akhrem, D. Mansuy, J. Fisch, R. Weiss and I. Morgenstern-

Badarau, J. Amer. Chem. Soc., **108** (1986) 5598-5607.

19. O. Wada, Y. Yano, G. Urata and K. Nakao, Biochem. Pharmacol., **17** (1968) 595-603.

20. M.D. Waterfield, A. Del Favero and C.H. Gray, Biochim. Biophys. Acta, **184** (1969) 470-473.

21. F. De Matteis, F.E.B.S. Letters, **6** (1970) 343-345.

22. F. De Matteis, Biochem. J., **124** (1971) 767-77.

23. F. De Matteis, Drug Metab. Disp., **1** (1973) 267-74.

24. W. Levin, E. Sernatinger, M. Jacobson, and R. Kuntzman, Science, **176** (1972) 134-43.

25. U.A. Meyer, H.S. Marver, Science, **171** (1971) 64-66.

26. S. Schwartz, K. Ikeda, Ciba Foundation Symposium on Porphyrin Biosynthesis and Metabolism, (1955) 209-28.

27. G. Abbritti and F. De Matteis, Chem. Biol. Interact., **4** (1971) 281-286.

28. W. Levin, M. Jacobson, R. Kuntzman, Arch. Biochem. Biophys., **148** (1972) 262-69.

29. W. Levin, E. Sernatinger, M. Jacobson and R. Kuntzman, Science, **176** (1972) 134-143.

30. W. Levin, M. Jacobson, E. Sernatinger, R. Kuntzman, Drug. Metab. Disp., **1** (1973) 275-285.

31. P.R. Ortiz de Montellano and B.A. Mico, Mol. Pharmacol., **255** (1980) 128-135.

32. J. Onisawa and R.F. Labbe, J. Biol. Chem., **238** (1963) 724-727.

33. A.C. Lockheed, J.H. Dagg and A. Goldberg, Brit. J. Derm., **79** (1967) 96-102.

34. G. Abbritti and F. De Matteis, Enzyme, **16** (1973) 196-202.

35. F. DeMatteis and A.H. Gibbs, Biochem. J., **146** (1975) 285-287.

36. G.D. Sweeney, D. Janigan, D. Mayaman and H. Lai, S. Afr. J. Lab. Clin. Med., **17** (1971) 68-72.

37. F. De Matteis and L. Cantoni, Biochem. J., **183** (1979) 99-103.

38. P.R. Ortiz de Montellano, K.L. Kunze and G.S. Yost, Biochem. Biophys. Res. Commun., **83** (1978) 132-137.

39. P.R. Ortiz de Montellano, R.A. Stearns and K.C. Langry, Molec. Pharmacol., **23** (1984) 310-317.

40. P.R. Ortiz de Montellano, B.A. Mico and G. Yost, Biochem. Biophys. Res. Commun., **83** (1978) 132-137.

41. P.R. Ortiz de Montellano, G.S. Yost, B.A. Mico, S.E. Dinizo, M.A. Correia, and H. Kambara, Arch. Biochem. Biophys., **197** (1979) 524-533.

42. P.R. Ortiz de Montellano and B.A. Mico, Arch. Biochem. Biophys., **206** (1981) 43-50.

43. J.M. Loosemore, G.N. Wogan, C. Walsh, J. Biol. Chem., **256** (1981) 8705-8712.

44. D.J. Waxman and C. Walsh, J. Biol. Chem., **257** (1982) 10446-10457.

45. L.M. Bornheim, A.N. Kotake and M.A. Correia, Biochem. J., **227** (1985) 277-286.

46. P.R. Ortiz de Montellano, H.S. Beilan, K.L. Kunze and B.A. Mico, J. Biol. Chem., **256** (1981) 4395-4399.

47. K.L. Kunze, B.L.K. Mangold, C. Wheeler, H.S. Beilan and P.R. Ortiz de Montellano, J. Bio. Chem., **258** (1983) 4202-4207.

48. P.R. Ortiz de Montellano, K.L. Kunze and B.A. Mico, Molec. Pharmacol., **18** (1980) 602-605.

49. P.R. Ortiz de Montellano, K.L. Kunze, H.S. Beilan and C. Wheeler, Biochem., **21** (1982) 1331-1339.

50. P.G. Gervasi, L. Citti, G. Fassina, E. Testai and G. Turchi, Toxicology Lett., **16** (1983) 217-223.

51. G. Abbritti and F. De Matteis, Chem. Biol. Interact., **4** (1972) 281-286.

52. F. De Matteis, A.H. Gibbs, L. Cantoni, J. Francis, Environmental Chemicals, Enzyme Function, and Human Disease, Ciba Found. Symp., **76** (1980) 119-139.

53. C. Ioannides, D.V. Parke, Chem. Biol. Interact., **14** (1976) 241-249.

54. P.G. Gervasi, L. Citti, G. Fassina, E. Testai and G. Turchi, Toxicol. Lett., **16** (1983) 217-223.

55. E. Testai, L. Citti, P. Gervasi, G. Turchi, <u>Biochem.</u>
    <u>Biophys.</u> <u>Res.</u> <u>Commun.</u>, **107** (1982) 633-641.

56. K.M. Ivanetich, S. Lucas, J.A. Marsh, M.R. Ziman, I.D.
    Katz, J.J. Bradshaw, <u>Drug.</u> <u>Metab.</u> <u>Disp.</u>, **6** (1978) 218-
    225.

57. J.M. Patel, E. Ortiz and K.C. Leibman, <u>Fed.</u> <u>Proc.</u>, **40**
    (1981) 636.

58. O. Augusto, K.L. Kunze and P.R. Ortiz de Montellano, <u>J.</u>
    <u>Bio.</u> <u>Chem.</u>, **257** (1982) 6231-6241.

59. P.R. Ortiz de Montellano, P.R. Beilan and K.L. Kunze,
    <u>Proc.</u> <u>Natl.</u> <u>Acad.</u> <u>Sci.</u>, <u>U.S.A.</u>, **78** (1981) 1490-1494.

60. P.R. Ortiz de Montellano, K.L. Kunze and H.S. Beilan, <u>J.</u>
    <u>Biol.</u> <u>Chem.</u>, **258** (1983) 45-47.

61. J.P. Collman, T. Kodadek, S.A. Raybuck, J.I. Brauman and
    L.M. Papazian, <u>J.</u> <u>Amer.</u> <u>Chem.</u> <u>Soc.</u>, **107** (1985) 4343-
    4345.

62. J.P. Collman, P.D. Hampton and J.I. Brauman, <u>J.</u> <u>Amer.</u>
    <u>Chem.</u> <u>Soc.</u>, **108** (1986) 7861-7862.

63. D. Mansuy, J.-P. Battioni, I. Akhrem, D. Dupre,
    J. Fischer, R. Weiss and I. Morgenstern-Badarau,
    <u>J.</u> <u>Amer.</u> <u>Chem.</u> <u>Soc.</u>, **106** (1984) 6112-6114.

64. P.R. Ortiz de Montellano, B.A. Mico, H.S. Beilan, K.L.
    Kunze, <u>Molecular</u> <u>Basis</u> <u>of</u> <u>Drug</u> <u>Action</u>, (1981) 151-166.

65. I.N.H. White, U. Muller-Eberhard, <u>Biochem.</u> <u>J.</u>, **166**
    (1977) 57-64.

66. J.M. Mathews, P.R. Ortiz de Montellano, <u>Fed.</u> <u>Proc.</u>, **41**
    (1982) 1541.

67. P.R. Ortiz de Montellano, H.S. Beilan and J.M. Mathews,
    <u>J.</u> <u>Med.</u> <u>Chem.</u>, **25** (1972) 1174-1179.

68. P.R. Ortiz de Montellano and K.L. Kunze, <u>J.</u> <u>Am.</u> <u>Chem.</u>
    <u>Soc.</u>, **102** (1980) 7373-7375.

69. P.R. Ortiz de Montellano, B.A. Mico, J.M. Mathews, K.L.
    Kunze, G.T. Miwa and A.Y.H. Lu, <u>Arch.</u> <u>Biochem.</u> <u>Biophys.</u>,
    210 (1981) 717-728.

70. P.R. Ortiz de Montellano, <u>J.</u> <u>Biol.</u> <u>Chem.</u>, **260** (1985)
    3330-3336.

71. P.R. Ortiz de Montellano and K.L. Kunze, <u>Biochem.</u>, **20**
    (1981) 7266-7271.

72. P.R. Ortiz de Montellano, K.L. Kunze, G.S. Yost and B.A. Mico, Proc. Natl. Acad. Sci., U.S.A., 76 (1979) 746-749.

73. P.R. Ortiz de Montellano and K.L. Kunze, J. Biol. Chem., 255 (1980) 5578-5585.

74. I.N.H. White, Biochem. J., 174 (1978) 853-861.

75. I.N.H. White, Biochem. Pharmacol., 29 (1980) 3253-3255.

76. I.N.H. White, Chem. Biol. Interact., 39 (1982) 231-243.

77. D. Mansuy, J. LeClaire, M. Fontecave and M. Momenteau, Biochem. Biophys. Res. Commun., 119 (1984) 319-325.

78. P.R. Ortiz de Montellano and K.L. Kunze, Biochem. Biophys. Commun., 94 (1980) 443-449.

79. W.H. Schaefer, T.M. Harris and F.P. Guengerich, Biochem., 24 (1985) 3254-3263.

80. C. Decker, K. Sugiyama, M. Underwood and M.A. Correia, Biochem. Biophys. Res. Commun., 136 (1986) 1162-1169.

81. D. Mansuy, Pure and Appl. Chem., 52 (1980) 681-690.

82. F. De Matteis, G. Abbritti and A.H. Gibbs, Biochem. J., 134 (1973) 717-727.

83. F. De Matteis, A.H. Jackson and S. Weerasinghe, FEBS Lett., 119 (1980) 109-112.

84. F. De Matteis, and A.H. Gibbs, Biochem. J., 187 (1980) 285-288.

85. T.R. Tephly, B.L. Coffman, G. Ingall, M.S. Abou Zeit-Har, H.M. Goff, H.D. Tabba and K.M. Smith, Arch. Biochem. and Biophys., 212 (1981) 120-126.

86. F. DeMatteis, A.H. Gibbs and T.R. Tephly, Biochem. J., 188 (1980) 145-152.

87. F. De Matteis, A.H. Gibbs, P.B. Farmer and J.H Lamb, FEBS Lett., 129 (1981) 328-331.

88. F. De Matteis, C. Holland, A.H. Gibbs, N. De Sa and M. Rizzardini, FEBS Lett., 145 (1982) 87-91.

89. F. DeMatteis, A.H. Gibbs and C. Hollands, J. Biochem., 211 (1983) 455-461.

90. G.S. Marks, D.T. Allen, C.T. Johnston, E.P. Sutherland, K. Nakatsu, and R.A. Whitney, Molec. Pharmacol., 27 (1985) 459-465.

91. K.L. Kunze and P.R. Ortiz de Montellano, J. Am. Chem. Soc., **103** (1981) 4225-4230.

92. D. Reichhart, A. Simon, F. Durst, J.M. Mathews and P.R. Ortiz de Montellano, Arch. Biochem. Biophys., **216** (1982) 522-529.

93. O. Augusto, H.S. Beilan and P.R. Ortiz de Montellano, J. Bio. Chem., **257** (1982) 11288-11295.

94. P.R. Ortiz de Montellano, H.S. Beilan and K.L. Kunze, J. Biol. Chem., **256** (1981) 6708-6710.

95. H.G. Jonen, J. Werringloer, R.A. Prough, R.W. Estabrook, J. Biol. Chem., **257** (1982) 4404-4411.

96. I.N.H. White, A.G. Smith and P.B. Farmer, Biochem. J., **212** (1983) 599-608.

97. P.R. Ortiz de Montellano, J.M. Mathews and K.G. Lanory, Tetrahedron, **40** (1984) 511-519.

98. R.A. Stearns and P.R. Ortiz de Montellano, J. Amer. Chem. Soc., **107** (1985) 234-240.

99. D.D.H. Craig, Lancet, **559** (1962).

100. N.R. Eade and K.W. Renton, J. Pharmacol. Exp. Therap., **173** (1970) 31.

101. S.F. Muakkassah, W.C.T. Yang, J. Pharm. Exp. Ther., **219** (1981) 147-155.

102. P.R. Ortiz de Montellano, O. Augusto, F. Viola, and K.L. Kunze, J. Biol. Chem., **258** (1983) 8623-8629.

103. P.R. Ortiz de Montellano and K.L. Kunze, J. Am. Chem. Soc., **103** (1981) 6534-6536.

104. S. Saito and H.A. Itano, Proc. Natl. Acad. Sci., U.S.A., **78** (1981) 5508-5512.

105. P.R. Ortiz de Montellano and D.E. Kerr, J. Biol. Chem., **258** (1983) 10558-10563.

106. P.R. Powell-Jackson, J.M. Tredger, H.M. Smith, M. Davis and R. Williams, Biochem. Pharmacol., **31** (1982) 4031-4034.

107. S. Hara, T. Satoh, and H. Kitagawa, Res. Commun. Chem. Path. Pharmacol., **36** (1982) 260-272.

108. S.F. Muakkassah, W.R. Bidlack, and W.C.T. Yang, Biochem. Pharmacol., **30** (1981) 1651-1658.

109. S.F. Muakkassah, W.R. Bidlack and W.C.T. Yang, <u>Biochem.</u>
     <u>Pharmacol.</u>, **31** (1982) 249-251.

110. R.N. Hines and R.A. Prough, <u>J. Pharmacol. Exp. Ther.</u>,
     214 (1980) 80-86.

111. P. Battioni, J.-P. Mahy, M. Delaforge and D. Mansuy,
     <u>Eur. J. Biochem.</u>, **134** (1983) 241-248.

112. T. Kawanishi, Y. Ohno, A. Takahashi,. Y. Kasuya and Y.
     Omuri, <u>Biochem. Pharmacol.</u>, **34** (1985) 919-924.

113. E.A. Smuckler, E. Arrhenius and J. Hultin, <u>Biochem. J.</u>,
     103 (1967) 55-64.

114. P.R. Ortiz de Montellano and K.L. Kunze, <u>Arch. Biochem.</u>
     <u>Biophys.</u>, **209** (1981) 710-712.

115. Ortiz de Montellano, P.R., Mathews, J.M., <u>Biochem. J.</u>,
     195 (1981) 761-764.

116. P.R. Ortiz de Montellano and J.M. Mathews, <u>Biochem.</u>
     <u>Pharmacol.</u>, **30** (1981) 1138-1141.

117. P.R. Ortiz de Montellano, B.A. Mico, J.M. Mathews, K.L.
     Kunze, G.T. Miwa and A.Y.H. Lu, <u>Arch. Biochem.</u>
     <u>Biophys.</u>, **210** (1981) 717-728.

118. L.B. Kier and E.B. Roche, <u>J. Pharm. Sci.</u>, **56** (1967)
     149-168.

119. F.H.C. Stewart, <u>Chem. Rev.</u>, **64** (1964) 129-147.

120. H. Wagner and J.B. Hill, <u>J. Med. Chem.</u>, **17** (1974)
     1337-1338.

121. J.B. Hill, R.E. Ray, H. Wagner and R.L. Aspinall,
     <u>J. Med. Chem.</u>, **18** (1975) 50-53.

122. R. Stejskal, M. Itabishi, J. Stanek and Z. Hruban,
     <u>Z. Virchows Arch. B.</u>, **18** (1975) 83-100.

123. P.R. Ortiz de Montellano and L.A. Grab, <u>J. Amer. Chem.</u>
     <u>Soc.</u>, **108** (1986) 5584-5589.

124. E.H. White and N. Egger, <u>J. Amer. Chem. Soc.</u>, **106**
     (1984) 3701-3703.

125. J.M. Patel, C. Harper and R.T. Drew, <u>Drug. Metab.</u>
     <u>Disp.</u>, **6** (1978) 368-378.

126. J.M. Patel, C.R. Wolf and R.M. Philpot, <u>Biochem.</u>
     <u>Pharmacol.</u>, **28** (1979) 2031-2036.

127. J.M. Patel, Toxicol. Appl. Pharmacol., **48** (1979) 337–342.

128. T.L. MacDonald, Z. Karimullah, L.T. Burka, P. Peyman, and F.P. Guengerich, J. Amer. Chem. Soc., **104** (1982) 2050–2052.

129. R.P. Hanzlik, R.H. Tullman, J. Am. Chem. Soc., **104** (1982) 2048–2050.

130. D. Mansuy and J.-P. Battioni, J. Chem. Soc., Chem. Commun., (1982) 638–639.

131. P.A. Krieter and R.A. Van Dyke, Chem. Biol. Interact., **44** (1983) 219–235.

132. O. Reiner and H. Uehleke, Hoppe-Seyler's Z. Physiol. Chem., **352** (1971) 1048.

133. H. de Groot, U. Harnisch and T. Noll, Biochem. Biophys. Res. Commun., **107** (1982) 885–891.

134. H.J. Ahr, L.J. King, W. Nastainczyk, and V. Ullrich, Biochem. Pharmacol., **31** (1982) 383–390.

135. D. Mansuy and M. Fontecave, Biochem. Pharmacol., **32** (1983) 1871–1879.

136. D. Mansuy, J.-P. Battioni and J.C. Chottard, J. Am. Chem. Soc., **100** (1978) 4311–4312.

137. M. Lange, and D. Mansuy, Tetrahedron Lett., **22** (1981) 2561–2564.

138. D. Mansuy, J.-P. Battioni D.K. Lavallee, J. Fischer and and R. Weiss, submitted for publication.

139. G. Fernandez, M.C. Villarruel, E.G. De Toranzo and J.A. Castro, Res. Chem. Commun. Pathol. Pharmacol., **35** (1982) 283–290.

140. M.J. Obrebska, P. Kentish, and D.V. Parke, Biochem. J., **188** (1980) 107–112.

141. G.E. Miller, M.A. Zemaitis and F.E. Greene, Biochem. Pharmcol., **32** (1983) 2433–2442.

142. J. Halpert, D. Hammond and R.A. Neal, J. Biol. Chem., **255** (1980) 1080–1089.

143. J. Leclaire, P.M. Dansette, D. Forstmeyer and D. Mansuy, Cytochrome P-450, Biochemistry Biophysics and Environmental Implications, Elsevier, NY, (1982) 95–98.

144. D. Mansuy, L. Devocelle, I. Artaud and J.-P. Battioni, Nouv. J. Chim., **9** (1985) 711-716.

145. J.T. Groves, R.C. Haushalter, M. Nakamura, T.E. Nemo and B. Evans, J. Amer. Chem. Soc., **103** (1981) 2884-2886.

146. J.P. Collman, T. Kodadek amd J.I. Brauman, J. Amer. Chem. Soc., **108** (1986) 2588-2592.

147. G.S. Marks, D.T. Allen, E.P. Sutherland, S.A. McCluskey and R.A. Whitney, Can. J. Physiol. Pharmacol., **64** (1986) 483-486.

148. I. Artaud, L. Devocelle, J.-P. Battioni, J.-P. Girault and D. Mansuy, J. Amer. Chem. Soc., **109** (1987) 3782-3783.

149. I. Artaud, N. Gregoire and D. Mansuy, submitted.

150. J.-P. Battioni, I. Artaud, D. Dupre, P. Leduc and D. Mansuy, Inorg. Chem., **26** (1987) 1788-1796.

151. T.G. Traylor and A.R. Miksztal, J. Amer. Chem. Soc., **109** (1987) 2770-2774.

152. T.G. Traylor, Y. Imamoto, B.E. Dunlap, A.R. Misztal and T. Nakano, J. Amer. Chem. Soc., **109** (1987).

CHAPTER 8

BIOCHEMISTRY OF N-SUBSTITUTED PORPHYRINS III: FORMATION BY REACTIONS OF HYDRAZINES WITH HEME PROTEINS AND MIGRATION REACTIONS OF MODEL COMPOUNDS

Introduction

Over a century ago, Hoppe-Seyler reported that a green pigment was formed in the erythrocytes of animals treated with phenylhydrazine.[1] This coloration is quite remarkable: administration of a few milligrams of phenylhydrazine to a rat causes the eyes to become green several minutes and the skin of the entire rat becomes green in about one-half hour. The phenylhydrazine causes the precipitation of erythrocytes in the form of Heinz bodies and, subsequently, hemolytic anemia. The formation of Heinz bodies and hemolytic anemia are also known to result from an inherited disorder in which an amino acid of hemoglobin is replaced, leading to altera-tion of the interaction of a heme with its neighboring amino acid residues.[2] The hemolysis which results from phenyl-hydrazine treatment of animals is similar in many respects to the hemolytic effects that are observed in the hereditary "Heinz-body anemias" (so-called because in 1890 R. Heinz observed aggregated particles in phenylhydrazine treated erythrocytes[3]). Induction of anemia in animals by phenyl-hydrazine treatment is the most common method for obtaining reticulocytes for biochemical studies. Webster found that the effects caused by the administration of phenylhydrazine and the formation of the green pigment were highly corre-lated.[4] Kiese and Seipelt showed that this green hemoglobin differs from verglobins formed by other reagents.[5] Beaven and White obtained the visible absorption spectrum of the isolated green product and remarked that the presence of a strong absorption band in the Soret region indicated an intact porphyrin ring,[6] not a type of biliverdin structure with an opened tetrapyrrole chromophore, as had previously been proposed.[7]

Although progress had been made in deducing the mechanism of Heinz body formation by phenylhydrazine,[8-11] both the mechanism by which the heme prosthetic group itself is altered and the structure of the green product remained unknown until work published by Saito and Itano[12] and Ortiz de Montellano and Kunze[13] in 1981 provided the definitive evidence.

Reactions of Aryl- and Alkylhydrazines with Myoglobin and Hemoglobin

First Reports of N-Phenylprotoporphyrin IX Formation. Saito and Itano reacted oxyhemoglobin and oxymyoglobin with stoichiometric amounts of phenylhydrazine hydrochloride and p-tolylhydrazinehydrochloride, respectively, in phosphate buffer (pH 7.4, 0.1 M) under aerobic conditions. After 1 h, solutions were treated with a mixture of acetic and hydrochloric acids for 20 h at $4^{\circ}C$. The products obtained by chloroform extraction were esterified with H_2SO_4 in methanol and separated chromatographically to give, in each case, a minor blue product and a major green product. Comparison of the ^1H nmr spectra and fragment patterns in the mass spectrum of non-phenyl-substituted analogs provided strong evidence that the blue products (formed in about 1/5 the amount of the green products) were σ-*meso*-phenylbiliverdin IXα methyl ester and σ-*meso*-p-tolylbiliverdin IXα methyl ester, respectively. (The parent compounds formed in the reactions of the arylhydrazines are, of course, the free acid forms rather than the dimethyl esters. Since the esters are much easier to purify chromatographically, the esterification step was used in their procedure.) The green (esterified) products were shown to have the same visible absorption spectrum as other N-substituted porphyrins (specifically, N-methylprotoporphyrin IX dimethyl ester[16]). The green pigments were converted to zinc(II) complexes (avoiding acid base equilibria of the uncomplexed free base with its mono-cationic or dicationic forms) to obtain ^{1}H nmr spectra. The nmr spectra are particularly informative. In addition to the shifts of the vinylic protons that had been found for N-methylprotoporphyrin IX dimethyl ester, the additional res-onances in the spectrum assignable to the N-phenyl protons are shifted greatly in the upfield direction: the ortho at 2.00 ppm, the meta at 5.01 ppm and the para at 5.57 ppm with respect to TMS. In addition, the shifts and broadening of the methylene protons of the propionic ester side chains led Saito and Itano to conclude that formation of the N-phenyl-protoporphyrin IX is specific, with substitution of the phenyl group occurring on the pyrrolenine rings bearing the propionic acid side chains.[12]

Saito and Itano proposed the following mechanisms for the formation of phenyl derivatives of biliverdin and proto-porphyrin. The formation of the blue products corresponding to aryl substituted biliverdins with the absence of unsub-stituted biliverdin was attributed to the close coupling of the oxidation and substitution reactions. This could occur, according to the authors, by reaction of oxyhemoglobin with phenylhydrazine to give a hydrogen peroxide–heme intermedi-ate, while rapid reaction of the phenyldiazene product with oxygen leads to production of a phenyl radical (Scheme 8.1). The hydrogen peroxo heme species, as previously proposed,[15] could lead to meso oxidation and porphyrin ring cleavage which would be accompanied by reaction of the phenyl radical at the σ-meso position. The formation of N-phenylproto-porphyrin was attributed to the attack of a phenyl radical on a porphyrin nitrogen atom followed by reduction of the product by a second molecule of phenylhydrazine to reduce the oxo-Fe(IV) or hydrogen peroxo N-phenyl heme interme-diate, producing phenyldiazene and preventing porphyrin ring cleavage. This mechanism accounted for the observation that two molecules of oxygen per heme are rapidly consumed in the reaction of an excess of phenylhydrazine with oxyhemoglo-bin.[15] Subsequent work has led to a revision of this mechanism (discussed later in the chapter), but this work was the first in which the mechanism proposed for this type of reaction was based on structural evidence.

$$Fe^{II}O_2 \; + \; C_6H_5NHNH_2 \longrightarrow [Fe^{II}H_2O_2] \; + \; C_6H_5N{:}NH$$

$$[Fe^{II}H_2O_2] \; + \; C_6H_5NHNH_2 \longrightarrow Fe^{II} \; + \; 2H_2O \; + \; C_6H_5N{:}NH$$

$$C_6H_5N{:}NH \; + \; O_2 \longrightarrow C_6H_5N{:}N{\cdot} \; + \; H^+ \; + \; O_2^-$$

$$C_6H_5N{:}N{\cdot} \longrightarrow C_6H_5{\cdot} \; + \; N_2$$

Scheme 8.1

In a report of concurrent work in his laboratory, Ortiz de Montellano presented the results of the identification of the product of the reaction of human hemoglobin with phenyl-hydrazine hydrochloride (in excess) under conditions similar to those used by Saito and Itano.[13] After workup in acidic methanol, the product was isolated by thin layer chromatography and high performance liquid chromatography. No blue products were reported. The HPLC procedure results in two fractions which, based on work with N-methylprotoporphyrin IX dimethyl esters,[14] one of which contains the two regio-isomers that have the N-phenyl group bound to the pyrrole-nine rings bearing propionic acid side chains and the other contains two regioisomers. The absorption spectra of these fractions are very similar to each other and to the unre-solved mixture reported by Saito and Itano. There is, as expected, only one parent mass (666, corresponding to the sum of masses of protoporphyrin IX dimethyl ester and a phenyl group.) Ortiz de Montellano and Kunze also converted the unresolved mixture to the zinc complex and obtained the ^{1}H nmr spectrum, assigning the additional peaks due to the N-phenyl ring in the same manner as Saito and Itano. From the results of their HPLC separation, it appears that both types of regioisomer are formed although the relative amounts were not reported. It is possible that the broadened methylene resonance of the propionic side chains in the nmr spectrum of the unresolved mixture is due to over-lapping peaks of the two types of isomer. These authors emphasized that the reaction requires aerobic conditions and that it occurs even with heme, eliminating the need for participation of the globin envelope in the analogous aryla-tion reactions of hemoglobins and myoglobin.[13] The possi-bility that N-phenylprotoporphyrin formation results from attack of a phenyl radical on the heme moiety was advanced, but with the reservation that further work was in progress to better define the mechanism.

Evidence from Kinetics and Visible Absorption Spectros-copy Concerning the Reaction Intermediate. In 1982, three groups published results of hydrazine reactions with hemoglobin. Ortiz de Montellano and coworkers found evidence for a protein-stabilized iron-phenyl intermediate,[17] Mansuy and coworkers compared the reactions of methyl- and phenyl-hydrazine and found evidence for an Fe(II) methyldiazene complex,[18] and Itano and Matteson determined the dependence of the rate of the reaction on the nature of substitutents on the phenyl ring of arylhydrazines,[19]

In their study, Ortiz de Montellano and coworkers mea-sured kinetic parameters, determined stoichiometries and isolated and identified the products of the reactions of hemoglobin, methemoglobin, myoglobin and metmyoglobin with phenyl-, 2-chlorophenyl, 3-chlorophenyl- 2-methylphenyl-, 3-methylphenyl- benzyl- and methylhyrazine. To establish the reaction stoichiometry, the yield of benzene from phenylhyd-razine, for example, was determined by gas-liquid chromato-graphy and the concentration of the heme protein from its absorbance in the visible region. In each case examined, the benzene/heme ratio was about five and the consumption of hydrazine was about six times that of the amount of heme present. Thus, at the termination of the catalysis of hydrazine oxidation, the fate of one of the molecules of hydrazine was not accounted for in terms of benzene produc-tion (or substituted benzenes in the cases of hydrazines other than phenylhydrazine). After the rapid initial consumption of oxygen, the amount of oxygen consumed, the amount of heme measureable by the pyridine hemochromogen assay method, and the spectrum of the heme product were all determined. The coincidence of changes in the latter two parameters with the consumption of oxygen led them to conclude that that hemoglobin or myoglobin catalyzed oxidation is terminated by chemical alteration of the heme prosthetic group.

To see how the nature of the hydrazine affects the reaction, the authors investigated the rates of reactions and the types of products formed with a variety of hydrazines. Reaction rate constants (Table 8.1) were calculated from initial rates, using a second order rate law (first order each in hemoprotein and hydrazine) based on the stoichiometric consumption of oxygen and hydrazine. They are very similar for all of the arylhydrazines studied (relative to phenylhydrazine, the reaction with 2-chlorophenylhydrazine is somewhat slower and that of 3-methylphenylhydrazine somewhat faster) whereas the reactions of benzyl- and ethylhydrazine are much slower. Also of interest is the finding that the products (after treatment with acidic methanol *under aerobic conditions*) of the reactions with phenyl-, 4-chlorophenyl-, 3-methylphenyl- and ethylhydrazine are N-substituted porphyrins (Table 8.1), whereas non-N-substituted products are found for the 2-chloro-, 2-methyl- and benzylhydrazine reactions. The fact that the rate constants are similar but the products different was attributed to the very reasonable possibility that the intermediate formed with these species may be too unstable to exist through the procedure for structural determination. In all cases, use of the spin-trapping reagent 5,5-dimethyl-1-pyrrolidine-N-oxide led to ESR signals characteristic of carbon radicals, indicating the presence of aryl or alkyl radicals derived from the hydrazines during the reaction. This result was consistent with evidence for reactive intermediates such as phenyl radicals and phenyldiazene[20-22] in the reaction of phenylhydrazine with oxyhemoglobin.

Table 8.1 Rates of Oxygen Consumption and The Nature of
Products Formed in Reactions of Hydrazines
With Methemoglobin[a]

Substrate (R in RNHNH₂)	O₂ Consumption k, $M^{-1}s^{-1}$	Type of Product (wavelength max.) m/e[b]
Phenyl-	55	N-arylporphyrin 667 (430,518,550,613,670)
2-chlorophenyl-	32	heme
4-chlorophenyl-	66	N-arylporphyrin 701/703 (427,520,550,617,672)
2-methylphenyl-	49	heme
3-methylphenyl-	76	N-arylporphyrin 681 (431,522,553,616,675)
Benzyl-	2.2	heme
Ethyl-	0.3	N-alkylporphyrin 619 (417,512,547,592,650)

a Ref. 17.
b The observed peaks correspond to monoprotonated molecular
ions (M^{+} + 1), as is common for N-substituted porphyrins.

Ortiz de Montellano and coworkers' experimental results
provided strong evidence for the authors' conclusion that
radicals formed during the catalytic oxidation of hydrazines
could inactivate the hemoprotein by binding to the iron atom
and, under acidic conditions in the presence of oxygen, that
the phenyl group could transfer to a pyrrolic nitrogen atom,
leading to an N-substituted porphyrin product.[17] In this
article, the direct analogy provided by model studies[23,24]
was cited. These and other model studies will be discussed
in the next section of this chapter.

Mansuy and coworkers investigated the reaction of methyl-hydrazine with hemoglobin and myoglobin[25] and compared their results with those of Saito and Itano's for reactions of phenylhydrazine.[12]   Under aerobic conditions (10 mol of O_2/mol of Hb), they found that methylhydrazine reacts with deoxymyoglobin or with metmyoglobin to form a species with the same visible spectrum as that obtained in the direct reaction of methyldiazene with deoxymyoglobin. The product they observed reacts immediately with CO to give the known MbFe(II)CO complex. Upon addition of one equivalent of ferricyanide to the intermediate (or the species immediately formed with methyldiazene), a new species is formed; it has a spectrum very similar to the product obtained from the reaction of phenylhydrazine with hemoglobin (before extraction – the σ-phenyliron complex). Very similar results were obtained with the hemoglobin reactions. Like Ortiz de Montellano, Mansuy and coworkers found that aerobic extraction and treatment with acid gave N-phenylproto-porphyrin from the reactions of phenylhydrazine (in about 40% yield) but no N-methylprotoporphyrin in the case of the methylhydrazine or methyldiazene reactions.

These results provided evidence for the reaction mechanism for phenylhydrazine and methylhydrazine reactions with hemoglobin and myoglobin shown in Scheme 8.2.[25]

$$HbFe(III) \xrightarrow{C_6H_5NHNH_2} B \xrightarrow{H^+,\ CH_3OH} \text{N-PhPP IX}$$

$$HbFe(III) \xrightarrow{CH_3NHNH_2} A \xrightarrow{Fe(CN)_6^{3-}} B$$

Scheme 8.2

where the structure of A as HbFe(II)(NH=NCH$_3$) is strongly supported by the results for the direct reaction of methyl-diazene. They did not propose a specific structure for the complexes B and B', but later results, especially involving model systems, establishes the structure of B as a σ-phenyl-iron(II) complex and the characteristic spectrum exhibited by B' indicates a similar structure, a σ-methyliron(II) complex. The much lower stability of alkyl as compared with aryl σ-iron(II) porphyrin complexes, as established by model studies (discussed later), is consistent with the fact that no N-methylprotoporphyrin is isolated upon acidic aerobic extraction.

In their article, Itano and Matteson reported the kinetics of phenylhydrazine with human oxyhemoglobin in the absence and presence of catalase, superoxide dismutase and other reagents; the effects of conducting the reactions under air, N$_2$ and CO were also reported.[19] As Ortiz de Montellano and coworkers found,[17] the rate of the reaction in air without added enzymes was not cleanly second order and only the approximately linear early segments of the absorbance change with time could be reasonably fit to a second order rate law. In the presence of superoxide dismu-tase and catalase, however, the plot of the reciprocal of oxyhemoglobin concentration vs. time was nicely linear and the rate itself was slower than in the absence of the en-zymes. The rates were not altered either in the presence or absence of the enzymes by the addition of hydroxyl radical and singlet oxygen scavengers (ethanol, tert-butyl alcohol, triethylamine or 1,4-diazabicyclo[2.2.2]octane), providing evidence that the changes in the visible absorption spectrum which follow the addition of phenylhydrazine to oxyhemoglo-bin are not due to oxidants directly derived from oxygen itself. The authors, therefore attributed the reaction to the direct bimolecular reaction of the phenylhydrazine and oxyhemoglobin molecules. The slower apparent rate in the presence of the enzymes was attributed to formation of methemoglobin by hydrogen peroxide and superoxide.

Spectral overlay data were presented which demonstrated that the product of the reaction in air was neither methemoglobin nor a methemoglobin-phenyldiazene complex, that the product of the reaction under N_2 was methemoglbin and that the product under CO was carbonmonoxyhemoglobin (about 70%). A number of possibilities for the intermediate formed under aerobic conditions were discussed and eliminated, leaving as a possibility the production of a phenyl radical that could react to form phenyl adducts of biliverdin and protoporphyrin.[12] Thus, they concluded the initial event in the oxidative degradation of hemoglobin by phenylhydrazine is the molecular reaction between the two.[19] A structure for the intermediate was not proposed.

Table 8.2 presents the data for a series of arylhydrazines in the presence of the two enzymes. Substitution in the ortho position greatly decreased the rate constant while halo or alkyl substitution in the meta or para positions increased the rate constant substantially. Itano and coworkers had previously shown that ortho groups decrease the severity of Heinz body hemolytic anemia[26] and that 2,6-dichlorophenylhydrazine does not cause hemolytic anemia,[27] consistent with the kinetic results and the deduction that the bimolecular reaction is essential in hemoglobin degradation. They had also previously found, however, that substitution of halo or alkyl groups in the meta or para positions did not increase the severity of hemolytic anemia[26] and they attributed the difference to the possibility that these substituents underwent slower adduct formation, offsetting the faster initial reaction.[19] Ortiz de Montellano's finding of very similar stoichiometries for the reactions of 2-chloro-, 4-chloro- and unsubstituted phenylhydrazine, but a different ability of these species to form N-phenylprotoporphyrin[17] leads to a different possibility, however. The ortho substituted species may be ineffective at producing Heinz bodies because they do not produce an N-substituted protoporphyrin, while the meta and para substituted phenylhydrazines may be of about the same effect because they produce about the same amount of N-substituted protoporphyrin. The effect of phenylhydrazines on Heinz body formation may be more a function of the amount of N-substituted porphyrin produced than the rate of production.

Table 8.2 Rate Constants for the Reaction of Hydrazines With Oxyhemoglobin in the Presence of Enzymes[a]

Substrate (R in $RC_6H_5NHNH_2$)	k, $M^{-1}s^{-1}$	Substrate (R in $RNHNH_2$)	k, $M^{-1}s^{-1}$
2-fluoro	21 ± 0	2-methyl	41 ± 1
3-fluoro	79 ± 8	3-methyl	184 ± 1
4-fluoro	168 ± 7	4-methyl	147 ± 2
2-chloro	25 ± 1	2-ethyl	57 ± 2
3-chloro	292 ± 14	3-ethyl	276 ± 6
4-chloro	553 ± 50	4-ethyl	187 ± 13
2-bromo	22 ± 1	2-isopropyl	53 ± 1
3-bromo	346 ± 13	4-isopropyl	106 ± 9
4-bromo	674 ± 21	2-tert-butyl	15 ± 1
2-iodo	22 ± 3	4-tert-butyl	40 ± 4
3-iodo	294 ± 4		
4-iodo	381 ± 51		
2,3-dichloro	53 ± 2		
2,4-dichloro	61 ± 3		
2,5-dichloro	24 ± 1		
2,6-dichloro	no reaction		
3,4-dichloro	1819 ± 31		
3,5-dichloro	245 ± 8		

a Ref. 19. Conditions: Equimolar oxyhemoglobin   and hydrazine, 0.1 M potassium phosphate buffer, pH 7.4, 25 °C, concentrations of superoxide dismutase and catalase each 3.0 x 10^{-7} M. For R = phenyl, k = 64 ± 4 $M^{-1}s^{-1}$ under these conditions.

**Determination of the Structure of the Reaction Interme-
diate by NMR spectroscopy and Crystallography.**   In 1983,
Kunze and Ortiz de Montellano definitively established the
nature of the intermediate in the reaction of p-tolylhydra-
zine with myoglobin using nmr spectroscopy.[28] New peaks
resulting from the reaction were detected at 47, 20 and -58
ppm in a 3:2:2 ratio, assignable to the p-methyl, meta and
ortho protons of a tolyl group bound directly to a low spin
Fe(II) atom. These peaks are not altered by the addition of
cyanide, indicating that the iron atom is not available for
binding by cyanide.   Phenylhydrazine treated myoglobin gives
peaks at 8, 18 and -55 ppm in the ratio 1:2:2.   The intermediate is extractable and gives a visible absorption spectrum
σ-phenylprotoporphinatoiron(II) and not that of N-phenyl-
protoporphinatoiron(II).   In 1983, Ortiz de Montellano and
coworkers isolated the products from the reaction of ethylhydrazine with human hemoglobin (the four regio isomers of
N-ethylprotoporphyrin IX), separated them by HPLC and
obtained the circular dichroism spectrum of the major
product (the isomer with the ethyl group on ring C).[29] This
species was obtained with a high enantiomeric excess, as
evidenced by the magnitude of the CD spectrum, demonstrating
the stereospecificity of the formation of the N-alkylporphyrin and, therefore, of the σ-phenyliron porphyrin intermediate.   Because the geometry of the hemoglobin molecule is
so well known, they were able to assign the absolute configuration to the N-ethylprotoporhyrin IX molecule (in the
dimethyl ester form as a result of isolation procedures).
The circular dichroism spectrum and absolute configuration
are shown in Figure 7.4.   It was this novel approach to the
assignment of absolute configuration that allowed Ortiz de
Montellano to determine the chiral orientation of the prosthetic heme in cytochrome P-450 discussed in the previous
chapter.

The coup de grace came in 1984 in the publication by
Ringe, Petsko, Kerr and Ortiz de Montellano of the structure
of the intermediate of the reaction of phenylhydrazine with
myoglobin at 1.5 A resolution, which clearly shows a phenyl

ring axially bound to the heme moiety in the protein.[30]   The
likely mechanism of the overall reaction to form N-phenylprotoporphyrin via the σ-phenyl intermediate and the basis
for assuming the authenticity of the spectral comparisons
arise from model studies to be discussed subsequently.

Reactions of Hydrazines with Other Hemoproteins

Reactions of hydrazines have also been carried out with
cytochrome P-450[31] and catalase[32] to determine whether the
intermediates and terminal products of the reactions bear a
relation to the corresponding species in the reactions of
myoglobin and hemoglobin.

Reactions with Catalase. Soon after his study of the
reactions of hydrazines with hemoglobin and myoglobin, Ortiz
de Montellano studied the reaction of bovine liver catalase
with phenylhydrazine and its dependence on the concentration
of hydrogen peroxide present.[32]   The catalase enzymes, along
with glutathione peroxidase, protect cells from hydrogen
peroxide associated not only with respiration, but also with
the metabolism of xenobiotics and drugs, of especial
importance in liver tissue. The hepatic enzyme which Ortiz
de Montellano and Kerr employed is very well characterized
and is a tetramer of four identical heme-containing units.
The effect of the presence of catalase on the reaction of
phenylhydrazine with hemoglobin was discussed above, but the
reaction of the catalase itself with phenylhydrazine had not
been reported before Ortiz de Montellano's study.

Addition of phenylhydrazine to catalase in the absence of
added hydrogen peroxide caused a very slow deactivation of
the enzyme and that only after a lag period. The lag period
and slow reaction were very reasonably attributed to a slow
increase in hydrogen peroxide concentration due to the production of the reaction of phenylhydrazine with the ferric
hemoprotein.

When hydrogen peroxide was added to solutions of phenylhydrazine and catalase, the rate at which deactivation occurred was directly proportional to the hydrogen peroxide concentration. The deactivation was irreversible, persisting after dialysis of the solutions of deactivated enzyme. These observations, coupled with the finding of benzene as a product, show that phenylhydrazine is a suicide substrate for catalase. Ortiz de Montellano and coworkers determined the approximate stoichiometry for the reaction to be:

$$3PhNHNH_2 + 52H_2O_2 + heme \longrightarrow 2PhH + 3N_2 + 54H_2O + 25O_2 + 3/2H_2 \\ + \sigma\text{-phenylheme}$$

where the coefficients of the products are inferred from the normal catalytic stoichiometry. Of the three phenylhydrazine molecules that react, one produces a phenyl-heme complex and the other two abstract a hydrogen atom to give benzene (the benzene production was determined qualitatively). In addressing the question of which of the generally postulated intermediates in the catalase cycle — Compound I, which is a two-electron oxidized intermediate, or Compound II, which is obtained when Compound I accepts an electron from a donor molecule — the authors pointed out that phenylhydrazine is the only one-electron donor in the reaction mixture and must react with Compound I. Either the phenyl radical produced or another molecule of phenylhydrazine could transfer an electron to Compound II to produce the ferric resting state of the enzyme. The phenyl radicals needed to form benzene and produce the σ-phenyliron intermediate could be produced either by the decomposition of the initial phenyl radical or from phenyldiazene, the two-electron oxidation product of phenylhydrazine, as shown in Scheme 8.3.[32] The latter possibility is given some support by the study of Mansuy and coworkers on the reaction of methyldiazene,[25] described previously, and by their subsequent study of the reaction of hydrazine and diazenes with cytochrome P450, discussed below.

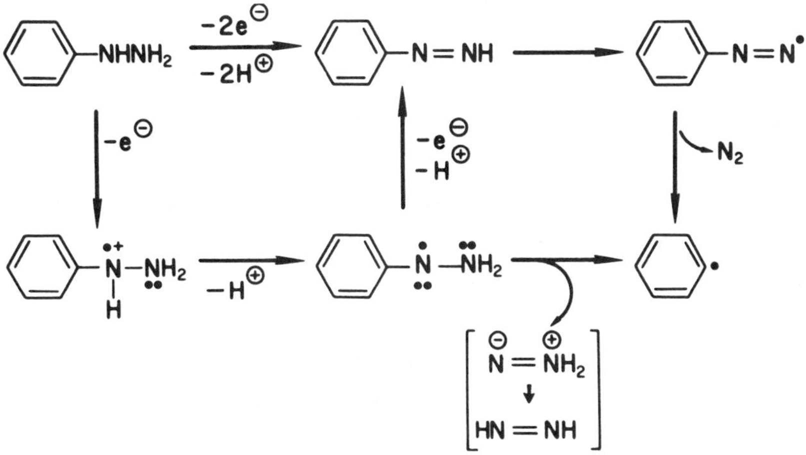

Scheme 8.3

Ortiz de Montellano and Kerr also reported the separation and identification of the products from the reaction of ethylhydrazine with catalase. After esterification, the porphyrin product was purified by HPLC into four isomeric fractions, with the isomers having the ethyl group bound to the C and D rings (bearing propionic acid side chains) predominant, as in the case of the reaction with hemoglobin. Although the C isomer is found in greatest abundance in both cases, the D isomer is more abundant for the reaction of catalase than of hemoglobin. Although alkylation of the C and D rings might be expected for hemoglobin (since this side of the heme ring lies near the entrance of the cavity leading to the iron atom) the authors remarked on the fact that this explanation is not so straightforward for catalase, based on its structure.[33] By the circular dichroism spectrum of the C isomer, which is the same as that found for hemoglobin (shown in Figure 7.4), they determined an orientation for the heme moiety in accord with the crystallographic results.

Catalase deactivation by phenylhydrazine is related to Heinz body formation and subsequent cell lysis is somewhat tenuous since, compared to hemoglobin, the reaction is slower and the concentration of catalase in erythrocytes is much lower. The mechanism advanced by the authors, with reservations as to its importance, is that the deactivation of catalase would greatly increase the vulnerability of the erythrocyte to hydrogen peroxide attack. If attack of hydrogen peroxide, or other radicals produced by reactions of hydrogen peroxide, on the erythrocyte membrane that leads to cell lysis, rather than direct effects of the precipitation of modified heme within the erythrocyte, the deactivation of catalase could be important.

Reactions with Cytochrome P-450 Enzymes. Mansuy and coworkers studied the reactions of hydrazines and diazenes with rat liver cytochrome P450 to determine if intermediates similar to those they observed for analogous reactions of hemoglobin were produced.[31] Since several monosubstituted hydrazines are used both industrially and medicinally, it is possible that their detoxification by hepatic cytochrome P450 may be of interest.

When alkyldiazenes are added to cytochrome P-450-Fe(II) under anaerobic conditions, the visible absorption spectrum of the enzyme changes within a few seconds, giving a new absorbance maximum at about 446 nm. The structure of this species was attributed to a complex derived from the simple addition of the diazene to the iron (II) atom. This conclusion was based upon the rapidity of formation, the similarity of the absorption maximum to those observed for ligands such as pyridines, imidazoles and primary amines, and the similarity of their reactivity with respect to CO (facile replacement), O_2 and dithionite (rapid reversion to the original ferrocytochrome P-450 spectrum) when compared with the complexes formed between diazenes and hemoglobin.[25] When these complexes are exposed to limited amounts of oxygen, a new species absorbing at about 480 nm is formed. Species with the same absorption spectrum are also produced by the direct addition of phenyldiazene to cytochrome P450-Fe(III) or by the oxidation of P450-Fe(III)-NH$_2$NHC$_6$H$_5$ by ferricyanide.[34] In their model studies using synthetic porphyrins, Mansuy and coworkers demonstrated that reactions of diazenes and hydrazines followed by oxidation lead unambiguously to σ-alkyl-iron(III) or σ-phenyl-iron(III) complexes.[35] By analogy, they assigned the structure of the 480 nm absorbing species derived from cytochrome P450 as a σ-alkyl- or σ-aryl-iron(III) hemoprotein. A important difference between the cytochrome derivatives and the species formed from synthetic porphyrins is the extremely large shifts of the Soret band. The authors mimicked this pronounced red shift by adding n-hexylthiolate to a model σ-phenyliron(III) porphyrin at -70°C (the data were not given, so it is not clear that the model porphyrin was the synthetic tetraphenylporphyrin or the naturally-derived protoporphyrin).[31] The overall scheme for the production of these species is given in Scheme 8.4.

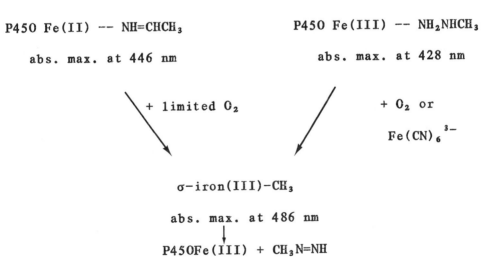

P450 Fe(II) -- NH=CHCH$_3$

abs. max. at 446 nm

+ limited O$_2$

P450 Fe(III) -- NH$_2$NHCH$_3$

abs. max. at 428 nm

+ O$_2$ or

Fe(CN)$_6$$^{3-}$

σ-iron(III)-CH$_3$

abs. max. at 486 nm

P450Fe(III) + CH$_3$N=NH

Scheme 8.4

A very important point made in this study is that the intermediate with the red-shifted Soret band near 480 nm derived from methyldiazene or methylhydrazine is extremely sensitive to oxygen, whereas the corresponding species derived from phenylhydrazine is moderately stable.[31] Thus, in the aerobic acidic workup used to isolate N-substitued porphyrins from hemoproteins, it would be expected that yields of N-phenylprotoporphyrin would be much higher than those of N-methylprotoporphyrin.

Model Studies Relevant to the Formation of N-Substituted Porphyrins in the Reactions of Hemoproteins with Hydrazines: Reactions of Iron Porphyrins

The spectral and structural characteristics of N-substituted porphyrins were investigated (because of the intrinsic interest of these unusual porphyrins[36-40]) before the biochemical characterization of these compounds. Similarly, much of the relevant model chemistry important in deducing the mechanism and reactivity patterns in the reactions of hydrazines with hemoproteins was developed earlier and, in fact, with metals other than iron. However, the explanations of the basis for the conclusions made in the case of the protein reactions is most clearly made by beginning with the reactions of iron. A later portion of this section will describe corresponding studies of the reactions of cobalt, nickel and zinc porphyrins chronologically. Most of these studies were made first because of the different stability of these metals with respect to reactions intermediates, provide interesting structural information that is not available for the iron complexes.

Evidence for the Formation of σ-Bound Organo-ironporphyrins Using Hydrazines. Definitive spectroscopic data establishing the structures of σ-aryl- and σ-alkyliron(III) intermediates, and the reversible formation of N-substituted porphyrins using tetraphenylporphinatoiron complexes as models, were reported in 1982 and 1983 by the groups of Ortiz de Montellano[41] and Mansuy.[18,35] Although a number of σ-alkyl iron porphyrin complexes had been reported before 1982,[42-47] Ortiz de Montellano's publication was the first in which a σ-phenyliron(III) porphyrin, specifically σ-phenyl-5,10,15,20-tetraphenylporphinatoiron(III), was synthesized and converted to the corresponding N-phenyl porphyrin.[41] The conditions used were like those used to isolate the N-substituted porphyrins from hemoproteins after hydrazine treatment: acidic methanolic solution in the presence of air. Exclusion of air led to a marked decrease in the yield of N-phenyltetraphenylporphyrin. The reactions sequence is given below (Scheme 8.5):

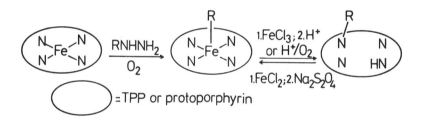

Scheme 8.5

As in the case of the intermediate formed from the reaction of myoglobin with phenylhydrazine (discussed previously) the nmr spectrum of σ-phenyltetraphenylporphyrin exhibits three new signals at widely separated chemical shifts (-25.52, 13.22 and -80.67 ppm with relative intensities of 1:2:2, assignable on the basis of selective irradiation experiments to the para, meta and ortho protons, respectively), with the ortho protons most shielded. The chemical shifts do not match those of the myoglobin intermediate closely, as might be expected from the facts that the protoporphyrin lies within the protein, the iron in the protein is coordinated to a histidyl nitrogen atom (as shown in the crystal structure[30]), and the porphyrin ligand is different. The characteristics of the nmr spectrum of the synthetic complex show that the complex is low spin, as also previously deduced for σ-alkyliron(III) porphyrins.[45]

Migration Reactions of σ-Bound Organic Moieties. In work carried out concurrently with that of Ortiz de Montellano, Mansuy and coworkers obtained conclusive spectroscopic evidence in tetraphenylporphinatoiron(III) for the migration of alkyl, aryl and vinyl group from the iron atom to the nitrogen atom under aerobic conditions to form the corresponding N-substituted porphyrins, including N-phenyltetraphenylporphyrin.[18] They synthesized σ-alkyl, aryl and vinyl complexes of Fe(III)tetraphenylporphyrin with methyl, $CH=C(C_6H_5)_2$ and phenyl as the axial ligands. Upon addition of ferric chloride (at $-20°$ C in toluene), the Fe(III) complexes were reduced, the axial group migrated to produce the corresponding chloro-N-substituted Fe(II) porphyrin complex. The visible-uv spectra of these N-substituted porphyrin complexes were very similar to those for other reported N-substituted porphyrin complexes and correspond exactly to the spectra for authentic samples they prepared by known methods. Using sodium dithionite as reducing agent, they reversed the migration: beginning with the Fe(II) complex of the N-substituted porphyrin, they produced the corresponding σ-bound Fe(III) porphyrin complex for each of the three organic groups. Addition of acid followed by base caused demetalation of the Fe(II) N-substituted porphyrin complexes, giving the free base N-substituted porphyrin, while addition of Fe(II) to the free base resulted in formation of the Fe(II) N-substituted porphyrin complex (which then formed the σ-bound complex upon reduction with dithionite), as shown in Scheme 8.6. This study demonstrated that under the appropriate conditions, the migration of a σ-bound organic moiety to form an N-substituted metalloporphyrin, with a corresponding change in the iron atom oxidation state from III to II, was a clean, well-defined reaction that occurred with a variety of organic groups. Although a good

deal of work type had been reported for cobalt complexes
(discussed below), this was the first report of its type for
iron porphyrin complexes. An interesting observation that
seems not to have been investigated extensively since this
publication was that the migration of the vinyl group, –
$CH=C(C_6H_5)_2$, also occurred during the resonance Raman study
of the complex, indicating that the iron(III)–carbon bond
could be activated with respect to migration photochemically
as well as by chemically.

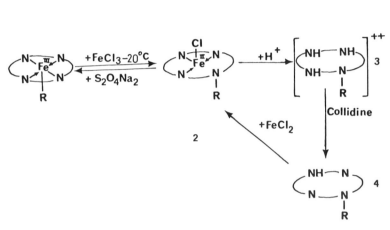

Scheme 8.6

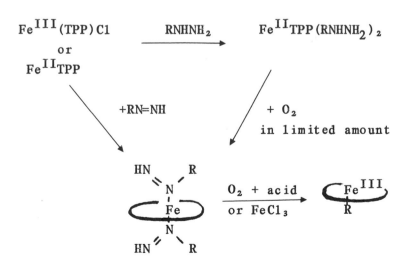

Scheme 8.7

In 1983, Mansuy and coworkers again used reactions of the synthetic Fe(III) tetraphenylporphyrin complex to shed light on biological reactions producing N-substituted porphyrins.[35] In this case, they reacted methylhydrazine, phenylhydrazine, methyldiazene and phenyldiazene with Fe(TPP)Cl in the presence of limited amounts of O_2 or $FeCl_3$, producing intermediates which were characterized spectrally as shown in Scheme 8.7. The observation of interest with respect to the reactions of hydrazines with hemoglobin is that the intermediate diazene complex is relatively stable in the case of the phenyl derivative but very unstable in the case of the methyl derivative. This finding is consistent with the fact that N-phenylprotoporphyrin is formed in relatively high yield from the reaction of phenylhydrazine with hemoglobin or myoglobin but N-methylprotoporphyrin is not.

Reactions of Carbene Iron Porphyrins. Also of interest are studies of iron(IV) carbene complexes. Mansuy and coworkers have made extensive studies of σ-bound carbene complexes, especially those derived from DDT,[48-50] and their work, as well as that of Balch and coworkers,[23] has demonstrated that a bridged carbene species, with the carbene moiety inserted between the pyrrolenine nitrogen atom and the iron atom, can be reversibly formed upon oxidation of the σ-carbene complex; oxidation under acidic conditions leads to the N-vinyl porphyrin,[50] as shown in Scheme 8.8.

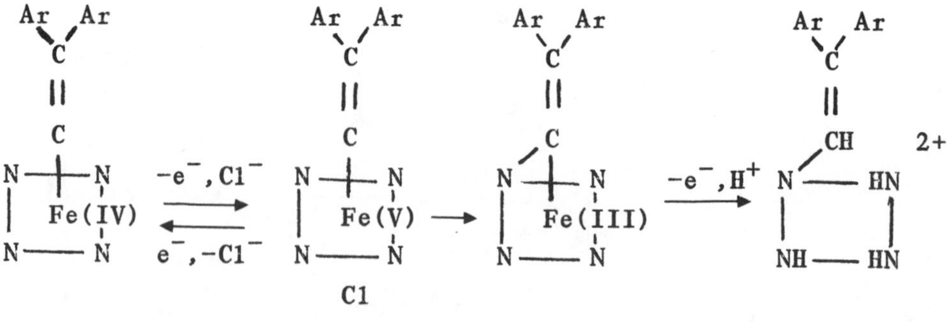

Scheme 8.8

Although such carbenes have only been isolated in a few favorable cases, their presence indicates that a structure with a carbene moiety between an iron atom and the pyrrolenine nitrogen may be at a minimum along a reaction path and that such structures may be intermediates in the production of N-substituted porphyrins *in vivo*.[51] The fact that these "inserted carbene" or metal—pyrrolenine bridged structures had also been found for other metals besides iron[52-55] (as discussed later) lends credence to the proposition that they may play a general role in metal-to-nitrogen migration reactions.

Model Studies Relevant to the Formation of N-Substituted Porphyrins in the Reactions of Hemoproteins with Hydrazines: Reactions of Cobalt, Nickel and Zinc Porphyrins

The earliest and most extensive literature with respect to the migration of σ-bound organic groups to form N-substituted porphyrins has involved cobalt, nickel and zinc complexes rather than those of iron. The reactions of the cobalt complexes are best characterized and are appear to be the most analogous to those of iron, although they were not the first studied.

Migration Reactions of σ-Bound Organocobalt Porphyrins. The first observation of the production of an N-alkyl porphyrin from a σ-alkylporphyrin complex of cobalt(III), specifically σ-methyloctaethylporphinatocobalt(III), was published by Ogoshi, *et al.*, in 1974.[56] Although the process and products were correctly interpreted, there was no supporting evidence for the mechanism of the reaction. More extensive work by Callot and coworkers demonstrated that σ-alkyl and σ-aryl cobalt(III) porphyrin complexes undergo intramolecular migration of the σ-bound moiety to form N-alkyl or N-aryl porphyrins in the presence of air.[57-60] In these articles the results were explicitly related to the production of N-alkylporphyrin "green pigments" *in vivo* from the interactions of drugs with cytochrome P-450 enzymes. They employed a wide variety of alkyl and aryl groups with cobalt(III) complexes of tetraphenylporphyrin and octaethylporphyrin and studied the effects of a variety of acids. Acetic acid is not strong enough to effect the migration, while acids such as hydrochloric, sulfuric and para-toluenesulfonic produce only XCo(III)porphyrins and porphyrin dications. The best conditions involve aerobic solutions of trifluoroacetic acid in dichloromethane, giving yields from 98% (for N-methylHTPP) to 28% (for N-methylOEP). They postulated a mechanism in which the small amounts of Co(III)porphyrin monocation that are present as a result of aerobic oxidation undergo a reversible migration of the axial organic ligand to form a Co(II) N-substituted porphyrin complex. Acidic demetallation renders the reaction irreversible and leads to the accumulation of N-substituted porphyrin (in a protonated form under the reaction conditions), as shown in Scheme 8.9.

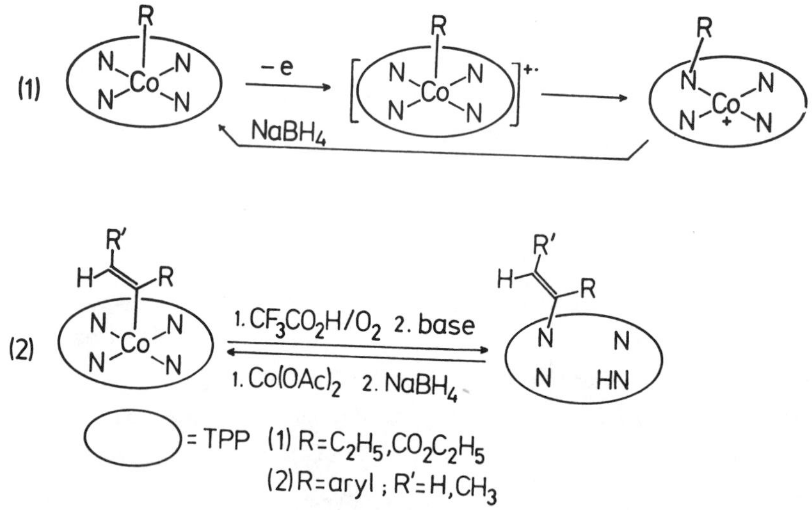

Scheme 8.9

Instead of using aerobic, acidic conditions to effect the migration, Callot used [(p-BrC$_6$H$_4$)$_4$N$^+$] SbCl$_6^-$, producing N-substituted complexes of cobalt(II). This result demonstrated the plausibility of the oxidative mechanism for these reactions. Further evidence for this mechanism comes from the work of Dolphin and his coworkers, who produced N-ethyltetraphenylporphyrin from σ-ethyltetraphenylporphinato-cobalt(III) and N-ethylacetoxytetraphenylporphyrin from σ-ethylacetoxytetraphenylporphinatocobalt(III) by electrochemical oxidation.[61] Callot and coworkers demonstrated that the migration reactions in both directions are intramolecular by using double labelling experiments in which either one or both of the σ-phenyl group of σ-phenyltetraphenylporphinatocobalt(III) and the four *meso*-phenyl groups were deuterated and the mass spectrum of the N-phenylHTPP product (or, for the reverse reaction, σ-phenyltetraphenylporphinatocobalt(III)) was observed. Lack of scrambling of the labels in this work was consistent with labelling experiments of Dolphin and his coworkers performed under electrochemical conditions, which also indicated that the migration to form N-substituted porphyrins is intramolecular. In the previous work, however, it was concluded that the reverse reaction is intermolecular. Callot ascribed the difference in results to thermal σ-bond cleavage and reformation in the solid state, based on the observation that the mass spectrum showed scrambling if the sample was heated slowly.[60]

The most thorough investigation of the mechanism of the migration of σ-bound organic ligands to form N-substituted porphyrins was the kinetic study of a series of N-aryl cobalt(III) complexes of tetraphenylporphyrin published by Callot and coworkers in 1984.[59] The migration reactions for the substituted phenyl groups were nearly quantitative, allowing them to obtain clean spectral overlays with well-defined isosbestic points. Reactions were carried out using one of three conditions: oxygen and trichloroacetic acid (which reacts at a more convenient, i.e., slower, rate than trifluoroacetic acid), $[(p-BrC_6H_4)_4N^+]$ $SbCl_6^-$, or electrochemically. In the case of the aerobic, acidic conditions, the overall reaction rate depends on the ease of oxidation of the starting material and they found a good correlation of the observed pseudo-first-order rate constant with the Hammett coefficients, σ_p^+, of the σ-aryl groups. (Rate constants ranged from 1.13 x 10^{-3} s^{-1} in the case of the p-methoxyphenyl complex to 8.2 x 10^{-6} s^{-1} for the p-nitrophenyl complex). Both the migration and demetallation steps (see Scheme 5.6) were found to be rapid. In the case of reactions with $[(p-BrC_6H_4)_4N^+]$ $SbCl_6^-$, the oxidation step is rapid, giving the visible absorption spectrum characteristic of a cation radical, while the migration step is rate determining and is highly sensitive to the nature of the aryl group (in the same order as the reactivity determined for the acidic, aerobic reactions). For the electrochemical reactions, an irreversible chemical step (the migration) occurred upon oxidation but rates were not determined. There was little effect of the σ-aryl group on the oxidation potential.

The reduction potential of the corresponding cobalt(II) complexes of the N-arylporphyrins is dependent on the N-aryl group. Since the bond between the N-substituted nitrogen atom and the cobalt(II) atom is relatively weak[62] and the differences between N-alkyl and N-aryl porphyrins appears to be due to electronic rather than structural effects,[63] this dependence seems most consistent with reduction of the porphyrin ring system rather than of the cobalt(II) atom (recognizing, of course, that the orbitals involved certainly have some character of each). Callot and coworkers found that the migration of the aryl group to the nitrogen atom (and loss of the chloride axial ligand) occur rapidly upon reduction and that the question of which step occurs first is moot at this point. They did find, however, that the calculated chemical rate is largely independent of the aryl group, indicating that little charge develops and that the migration is rate determining.

Besides the fact that these reactions provided the basis for the interpretation of the reactions of iron porphyrins in biological systems, they also provide a very practical means for the synthesis of N-substituted complexes without the necessity of resorting to highly reactive (and dangerous) alkylating agents. Because the reactions are intramolecular, they also provide a means of synthesizing N-substituted porphyrins having peripheral substituents that might be affected under the more vigorous conditions necessary for forming N-alkyl porphyrins by other means.

Formation of N-Substituted Products from Reactions of Diazenes with Zinc(II) Porphyrins.   Reactions of zinc porphyrins to form N-substituted porphyrins have been observed since 1972.[64]   Because σ-alkyl and σ-aryl zinc metalloporphyrins are not as stable as corresponding cobalt(III) complexes, the nature of the intermediates and the mechanisms of their reactions has not been as straightforward to interpret. The route by which N-substituted porphyrins have been produced from zinc porphyrins (either tetraphenylporphyrin or octaethylporphyrin) involves the reaction of a diazoester through an intermediate, postulated to be a zinc-alcoxycarbonylalkyl porphyrin.[65-68]   The formation of an N-alkylporphyrin product appears to require acid (in some cases provided by water[68]).   Since all of these reactions reported with zinc complexes have been done under aerobic conditions, it is likely that the mechanism is similar to that which has been well established for cobalt(III) complexes, as shown in Scheme 8.10.   It should be noted that only the initial zinc complex and the N-alkylporphyrin product have been identified and that the two intermediates are postulated based on the cobalt(III) reactions. It is also possible that a carbene intermediate is formed before the N,N' bridged intermediate,[68] as in the reactions of Fe(III) porphyrins.[24] It should also be noted that the reactions of zinc(II) porphyrin complexes with diazoacetates leads, especially in the absence of a source of protons, to the formation of

homoporphyrins, in which a carbene moiety is added to a
methine position to give an expanded ring system.[65,66,68]
These reactions and the mechanism were discussed previously
(Chapter 4).

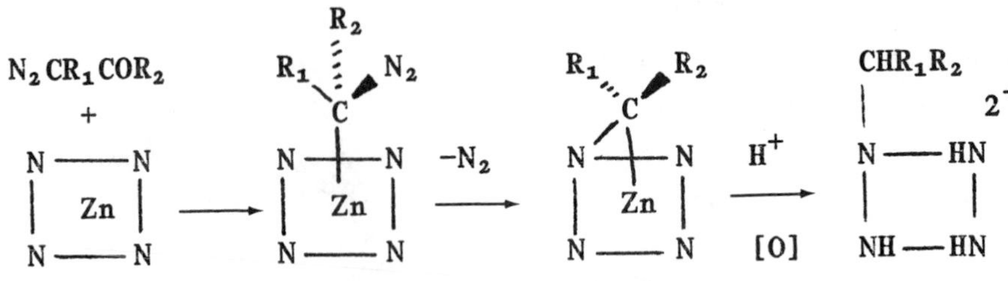

Scheme 8.10

To convert the N-acetoxy porphyrins to N-alkylporphyrins,
Callot and coworkers converted the esters to the acid form
which readily undergoes photolytic decarbonylation, with the
zinc porphyrin acting as its own photosensitizer.[67] This
reaction is especially efficacious for the formation of N-
methylporphyrins. The reaction sequence is illustrated
below (Scheme 8.11).

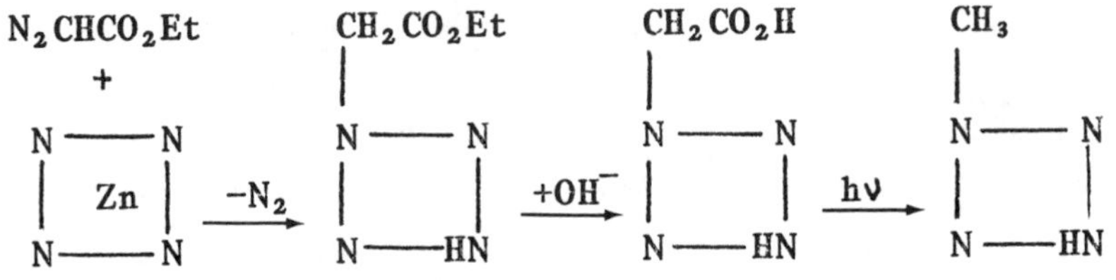

Scheme 8.11

Formation of N-ethyltetraphenylporphyrin from methyl
diazopropionate or, better, methyl pyruvate tosylhydrazone,
does not give as high a yield (25% vs. 53%). The N-CH_3HTPP

can also be obtained in good yield by thermal decomposition of the acid form (67% yield based on the acid, 44% overall), but the $N-C_2H_5HTPP$ yield is lower (25% from the acid, 12% overall), which Callot and coworkers attributed to an elimination reaction producing acrylic acid, as shown in Scheme 8.12.[67]

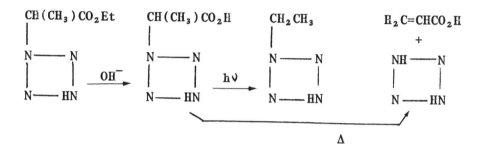

Scheme 8.12

Reactions of Rhodium(III) Porphyrins. It might be assumed that the rhodium(III) porphyrins, being congeners of cobalt(III) porphyrins, would have similar reactivity patterns with respect to formation of N-alkyl porphyrins. However, Callot and Schaeffer found that the bridged intermediate could not be isolated, that olefins were formed with diazopropionates, leading to $Rh-CO_2$-alkyl derivatives, and that the rhodium(III) porphyrins cause catalytic decomposition of diazoesters.[68] The major product (90% in the case of OEP) of the addition of ethyl diazoacetate to the methylrhodium(III) complexes of octaethyl- or tetraphenylporphyrin were characterized as the σ-methyl-N-ethylacetoxy rhodium(III) porphyrin complexes. The predominant products of reactions of the iodorhodium(III) porphyrins are the σ-acetoxyrhodium(III) porphyrins, but some "green products" were also formed, presumed to have N-acetoxy substituents. A comparison of the reaction paths for cobalt(III) and rhodium(III) porphyrins is given in Scheme 8.13. Formation

of the N-acetoxy products of the rhodium porphyrins requires
a second alkylation step. This subsequent alkylation is
much more favorable if there is a σ-bound methyl group is
present.

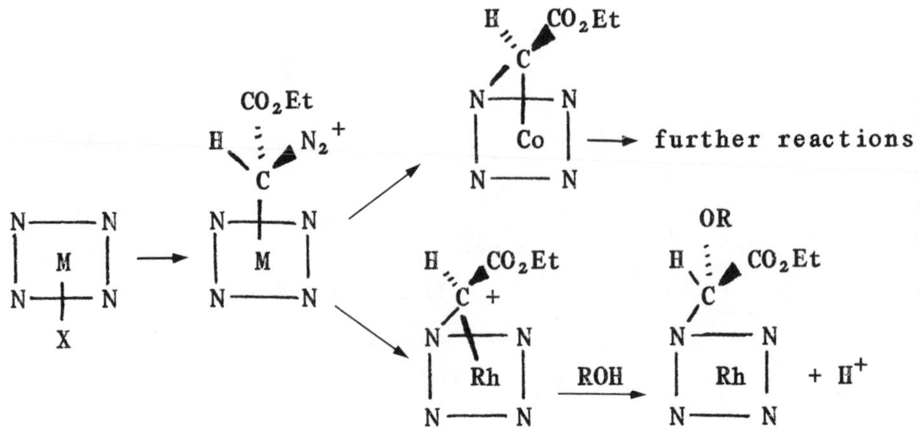

Scheme 8.13

Reactions of rhodium(I) complexes of octaethylporphyrin
and etioporphyrin, as well as corroles had been studied
previously, but the reactions were carried out in the
opposite sense. That is, the initial species used was the
N-alkyloctaethylporphyrin[69,70] N-alkyletioporphyrin[71] or
corrole[72] and the object of the studies was to observe the
reaction which occurs upon addition of rhodium(I) (as di-μ-
bis(dicarbonylrhodium)). In the cases of the porphyrins,
the products are the corresponding σ-alkylrhodium(III)
porphyrins. (Similar results are obtained using N-methyloctaethylporphyrin and iridium(I)[73]).

In a very interesting experiment, Grigg and coworkers found that if the reaction of N-methyletioporphyrin with di-μ-chloro-bis(dicarbonylrhodium) is carried out in the presence of an excess of ethyl iodide, the product is about a 1:1 mixture of σ-methylrhodium(III)etioporphyrin and σ-ethylrhodium(III)etioporphyrin.[72] In the case of the corroles, the N,N'-di-carbonylrhodium(I), N''methyl corrole complexes are relatively stable and have been characterized by x-ray crystallography.[72]

Grigg and coworkers found that the presence of the di-μ-chloro-bis(carbonylrhodium) moiety bound to a porphyrin promotes incorporation of an alkylating agent from an alkyl halide precursor, resulting in an alkylrhodium(III) porphyrin product. Therefore, it is possible that N-alkyl-porphyrins are formed as intermediates in the reactions they reported between methyl- and ethyliodide and porphyrin and azaporphyrin (in which one of the methine carbon atoms and its hydrogen atom are replaced by a nitrogen atom) complexes of di-μ-chloro-bis(carbonylrhodium). These reactions lead to the formation of a mixture of alkyl-rhodium(III) and acyl-rhodium(III) products (where the acyl product is $-CO_2CH_3$ or $-CO_2C_2H_5$, respectively).[74] Since oxidative addition of a methyl group to a carbonyl ligand bound to rhodium(I) is known, however, to give both methylrhodium(III) and acetylrhodium(III) products, the products could also result from a direct attack on the carbonyl rather than on the pyrrolic nitrogen atom.

These intriguing results were reported a number of years ago and have not been extensively pursued even though the mechanistic questions are unsettled and the reactions themselves quite novel.

Reactions of Nickel and Palladium Porphyrins to Form Intermediates Implicated in Diazene Reactions. The reactions of nickel and palladium N-alkylporphyrin and N-alkylcorrole complexes constitute some of the earliest literature in the field of N-substituted porphyrin chemistry.[42,52,75-78] For the most part, these studies dealt with reactions in which the complex was formed by the direct alkylation of a porphyrin or corrole complex. In some cases, however, reactions of the nickel complexes of N-alkylporphyrins provided intermediates more stable than those of cobalt, iron or zinc which thus provide valuable (if not irrefutable) mechanistic evidence for the general pathway of migration reactions.

In 1974, Callot and Tschamber reported that N-ethylacetoxytetraphenylporphyrin reactions with bis(acetoxyacetone)nickel(II), by way of the N-ethylacetoxytetraphenylporphinatonickel(II) complex, to form a "homoporphyrin" in which the porphine ring system is expanded by addition of a carbon between one of the pyrrolenine rings and a methene carbon atom, as shown in Figure 8.1.[77] Further study revealed that the reaction proceeded via a bridged intermediate in which the $C(H)CO_2C_2H_5$ moiety was introduced between a pyrrolenine nitrogen atom and the nickel(II) atom[54,78] (the structure of this species[79] is discussed in some detail in Chapter 2).

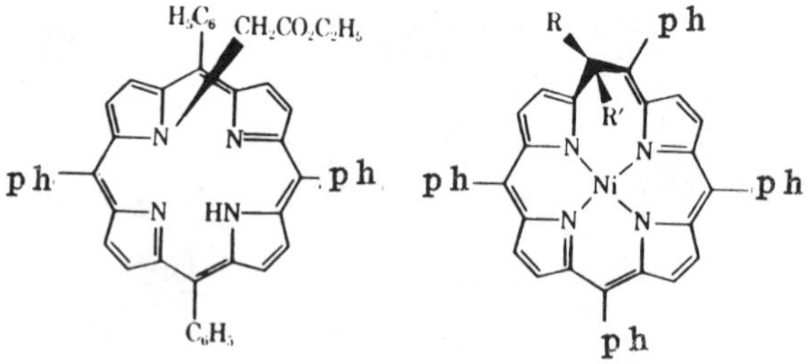

Figure 8.1 The structures of N-ethylacetoxyporphyrin (left) and the homoporphyrin product resulting from treatment with $Ni(acac)_2$ (right, the yield is 57% endo, 4% exo, ref. 78).

An Overall Mechanism for Reactions of Diazenes with Metalloporphyrins. From the combination of the structures isolated for the nickel reactions and those of cobalt and zinc, Callot proposed an overall mechanistic scheme for the migration reactions of N-substituted metalloporphyrins upon treatment with diazoalkanes[677] as shown in Scheme 8.15. This scheme also corresponds to the later observations of the reactions of iron porphyrins, including the reactions of the natural hemoproteins discussed earlier in this chapter.

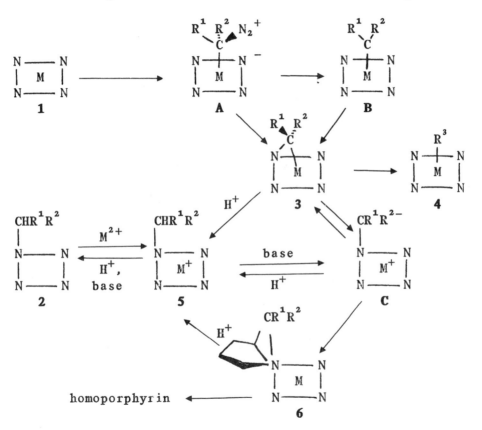

Scheme 8.15

The numbered species in this scheme have all been definitively characterized crystallographically and/or spectroscopically. Well-defined species are: **1**, Fe(III), Co(III), Zn(II), Rh(I) (as a dicarbonyl species) and Ni(II); **2**, Fe(II); **3**, Ni(II); **4**, Fe(III), Co(III), Rh(III); **5**, Fe(II), Co(II), Zn(II), Ni(II) and **6**, Ni(II).

Conclusion

Certainly, as Callot has pointed out,[67] the chemistry of the pyrrolenine nitrogen atoms of metalloporphyrins is more varied and interesting than had been thought fifteen years ago. The ability of an organic ligand to transfer from metal atom to nitrogen atom, often reversibly, via a formal oxidation-reduction reaction, driven by the greater stability of the low oxidation state of metal atoms in N-substituted complexes and higher oxidation states in non-N-substituted metalloporphyrins, is remarkable. The fact that this reaction is directly relevant to the natural decomposition of the cytochrome P-450 enzymes and the destruction of hemoproteins in the long-observed reactions of hydrazines is, of course, all the more interesting.

References

1. G.Z. Hoppe-Seyler, Physiol. Chem., 9 (1885) 34-39.

2. R.W. Carrell and H. Lehman, Semin. Hematol., 6 (1969) 116-132.

3. R. Heinz, Virchows Arch. Pathol. Anat. Physiol. Kln. Med., 122 (1890) 112-116.

4. S.H. Webster, Blood, 4 (1949) 479-497.

5. M.Keise and L. Seipelt, Naunyn-Schmeidebergs Arch. Exp. Pathol. Pharmakol., 200 (1943) 648-683.

6. G.H. Beaven and J.C. White, Nature (London), 173 (1954) 389-391.

7. R. Lemberg and J.W. Legge, Aust. J. Exp. Biol. Med. Sci., 20 (1942) 65-68.

8. J.K. French, C.C. Winterbourn and R.W. Carrell, Biochem. J., 173 (1978) 19-26.

9. B. Goldberg and A. Stern, Arch. Biochem. Biophys., 178 (1977) 218-225.

10. H.A. Itano, K. Hirota and T.S. Vedrick, Proc. Natl. Acad. U.S.A., 74 (1977) 2556-2560.

11. J. Peisach, W.E. Blumberg and E.A. Rachmilewitz, Biochim. Biophys. Acta, **393** (1975) 404-418.

12. S. Saito and H.A. Itano, Proc. Natl. Acad. Sci., U.S.A., **78** (1981) 5508-5512.

13. P.R. Ortiz de Montellano and K.L. Kunze, J. Am. Chem. Soc., **103** (1981) 6534-6536.

14. K.L. Kunze and P.R. Ortiz de Montellano, J. Am. Chem. Soc., **103** (1981) 4225-4230.

15. R. Bonnett and M.J. Dinsdale, J. Chem. Soc., Perkin Trans. I, (1972) 2540-2548.

16. T.S. Vedrick and H.A. Itano, Biochim. Biophys. Acta., **672** (1981) 214-218.

17. O. Augusto, K.L. Kunze and P.R. Ortiz de Montellano, J. Bio. Chem., **257** (1982) 6231-6241.

18. D. Mansuy, J.-P. Battioni, D. Dupre, E. Sartori and G. Chottard, J. Am. Chem. Soc., **104** (1982) 6159-6161.

19. H.A. Itano and J.L. Matteson, Biochem., **21** (1982) 2421-2426.

20. H.A. Itano and S. Mannen, Biochim. Biophys. Acta, **421** (1976) 87-96.

21. H.P. Misre and I. Fridovich, Biochem., **15** (1976) 681-687.

22. H.A.O. Hill and P.J. Thornalley, FEBS Lett., **125** (1981) 235-238.

23. L. Latos-Grazynski, R.-J. Gheng, G.N. La Mar and A.L. Balch, J. Am. Chem. Soc., **103** (1981) 4270-4272.

24. M. Lange, and D. Mansuy, Tetrahedron Lett., **22** (1981) 2561-2564.

25. D. Mansuy, P. Battioni, J.-P. Mahy and G. Gillet, Biochem. Biophys. Res. Commun., **106** (1982) 30-36.

26. H.A. Itano, K. Hosokawa and K. Hirota, Brit. J. Hematol., **32** (1976) 99-104.

27. H.A. Itano, K. Hirota and K. Hosokawa, Nature (London) **256** (1975) 665-667.

28. K.L. Kunze and P.R. Ortiz de Montellano, J. Am. Chem. Soc., **105** (1983) 1380-1381.

29. P.R. Ortiz de Montellano, K.L. Kunze and H.S. Beilan, <u>J.</u> <u>Biol.</u> <u>Chem.</u>, **258** (1983) 45-47.

30. D. Ringe, G.A. Petsko, D.E. Kerr and P.R. Ortiz de Montellano, <u>Biochem.</u>, **23** (1984) 2-4.

31. P. Battioni, J.-P. Mahy, M. Delaforge and D. Mansuy, <u>Eur.</u> <u>J.</u> <u>Biochem.</u>, **134** (1983) 241-248.

32. P.R. Ortiz de Montellano and D.E. Kerr, <u>J.</u> <u>Bio.</u> <u>Chem.</u>, **258** (1983) 10558-10563.

33. T.J. Reid, M.R.N. Murthy, A. Sicignano, H. Tanaka, M.D.L. Musick and M.G. Rossman, <u>Proc.</u> <u>Natl.</u> <u>Acad.</u> <u>Sci.,</u> <u>U.S.A.</u>, **78** (1981) 4767-4771.

34. H.G. Jonen, J. Werringloer, R.A. Prough, R.W. Estabrook, <u>J.</u> <u>Biol.</u> <u>Chem.</u>, **257** (1982) 4404-4411.

35. P. Battioni, J.P. Mahy, G. Gillet and D. Mansuy, <u>J.</u> <u>Am.</u> <u>Chem.</u> <u>Soc.</u>, **105** (1983) 1399-1401.

36. W.S. Caughey and P.K. Iber, <u>J.</u> <u>Org.</u> <u>Chem.</u>, **28** (1963) 269-270.

37. A.H. Jackson and G.R. Dearden, <u>Ann.</u> <u>New</u> <u>York</u> <u>Acad.</u> <u>Sci.</u>, **206** (1973) 151-176.

38. D.K. Lavallee and M.J. Bain, <u>Bioinorg.</u> <u>Chem.</u>, **9** (1978) 311-321.

39. D.K. Lavallee, O.P. Anderson and A. Kopelove, <u>J.</u> <u>Am.</u> <u>Chem.</u> <u>Soc.</u>, **100** (1978) 3025-3033.

40. D.K. Lavallee, <u>Inorg.</u> <u>Chem.</u>, **17** (1978) 231-236.

41. P.R. Ortiz de Montellano, K.L. Kunze and O. Augusto, <u>J.</u> <u>Am.</u> <u>Chem.</u> <u>Soc.</u>, **104** (1982) 3545-3546.

42. D.A. Clarke, R. Grigg and A.W. Johnson, <u>J.</u> <u>Chem.</u> <u>Soc.,</u> <u>Chem.</u> <u>Commun.</u>, (1966) 208-209.

43. D.A. Clarke, D. Dolphin, R. Grigg, A.W. Johnson and H.A. Pinnock, <u>J.</u> <u>Chem.</u> <u>Soc.</u>, (1968) 881-885.

44. D. Mansuy, <u>Pure</u> <u>and</u> <u>Appl.</u> <u>Chem.</u>, **52** (1980) 681-690.

45. D. Lexa, J. Mispelter and J.-M. Saveant, <u>J.</u> <u>Am.</u> <u>Chem.</u> <u>Soc.</u>, **103** (1981) 6806-6812.

46. P. Cocolios, E. Laviron and R. Guilard, <u>J. Organomet. Chem.</u>, **228** (1982) C39-C42.

47. H. Ogoshi, S. Kitamura, H. Toi and Y. Aoyama, <u>Chem. Soc. Jpn., Chem. Lett.</u>, (1982) 495-498.

48. D. Mansuy, M. Lange, J.C. Chottard, J.F. Bartoli, B. Chevrier and R. Weiss, <u>Angew. Chem.</u>, **90** (1978) 828-929.

49. D. Mansuy, M. Lange and J.C. Chottard, <u>J. Am. Chem. Soc.</u>, **100** (1978) 3213-3214.

50. D. Mansuy, J.P. Battioni, J.C. Chottard and V. Ullrich, <u>J. Am. Chem. Soc.</u>, **101** (1979) 3971-3973.

51. A.W. Johnson, D. Ward, P. Battan, A.L. Hamilton, G. Shelton and C.M. Elson, <u>J. Chem. Soc., Perkin I</u>, (1975) 2076-2085.

52. A.W. Johnson and D. Ward, <u>J. Chem. Soc., Perkin I</u>, (1977) 720-723.

53. P. Batten, A.L. Hamilton, A. W. Johnson, M. Mahendram, D. Ward and T.J. King, <u>J. Chem. Soc., Perkin I</u>, (1977) 1623-1628.

54. H.J. Callot, Th. Tschamber, B. Chevrier and R. Weiss, <u>Angew. Chem.</u>, **87** (1975) 545.

55. H. Ogoshi, E.I. Watanabe, N. Koketzu and Z.I. Yoshida, <u>J. Chem. Soc., Chem. Commun.</u>, (1974) 943-944.

56. H.J. Callot and E. Schaeffer, <u>Tetrahedron Lett.</u>, **21** (1980) 1335-1338.

57. H.J. Callot and E. Schaeffer, <u>J. Organomet. Chem.</u>, **193** (1980) 111-115.

58. H.J. Callot and F. Metz, <u>J. Chem. Soc., Chem. Commun.</u>, (1982) 947-948.

59. H.J. Callot, F. Metz and R. Cromer, <u>Nouv. J. Chim.</u>, **8** (1984) 759-763.

60. D. Dolphin, D.J. Halko and E. Johnson, <u>Inorg. Chem.</u>, **20** (1981) 4348-4351.

61. H.J. Callot and R. Cromer, <u>Nouv. J. Chim.</u>, **8** (1984) 765-770.

62. O.P. Anderson and D.K. Lavallee, <u>J. Amer. Chem. Soc.</u>, **98** (1976) 4670-4671.

63. D. Kuila, D.K. Lavallee, C.K. Schauer and O.P. Anderson, J. Am. Chem. Soc., **106**, (1984) 448-450.

64. H.J. Callot, Bull. Soc. Chim. Fr., **11** (1972) 4387-4391.

65. H.J. Callot and Th. Tschamber, Bull. Soc. Chim. Fr., **11** (1973) 3192-3198.

66. H.J. Callot, Tetrahedron Lett., **33** (1979) 3093-3096.

67. H.J. Callot and E. Schaeffer, Nouv. J. Chim., **4** (1980) 307-309.

68. H.J. Callot and E. Schaeffer, Nouv. J. Chim., **4** (1980) 311-314.

69. H. Ogoshi, T. Omura and Z. Yoshida, J. Amer. Chem. Soc., **95** (1973) 1666-1668.

70. H. Ogoshi, J. Setsune, T. Omura and Z. Yoshida, J. Amer. Chem. Soc., **97** (1975) 6461-6466.

71. A.M. Abeysekera, R. Grigg, J. Trocha-Grimshaw and K. Hendrick, Tetrahedron, **36** (1980) 1857-1868.

72. A.M. Abeysekera, R. Grigg and J.Trocha-Grimshaw, J. Chem. Soc., Perkin I, (1979) 2184-2192.

73. H. Ogoshi, J. Setsune and Z. Yoshida, J. Organomet. Chem., **159** (1978) 317-328.

74. A.M. Abeysekera, R. Grigg, J. Trocha-Grimshaw and V. Viswanatha, J. Chem. Soc., Perkin I, (1977) 36-44.

75. R. Grigg, A.W. Johnson and G. Shelton, J. Chem. Soc., (1971) 2287-2294.

76. R. Grigg, G. Shelton and A. Sweeney, J.Chem. Soc., Perkin Trans. I, (1972) 1789-1799.

77. H.J. Callot and Th. Tschamber, Tetrahedron Lett., **36** (1974) 3155-3158.

78. H.J. Callot and Th. Tschamber, J. Amer. Chem. Soc., **97** (1975) 6175-6178.

79. B. Chevrier and R. Weiss, J. Am. Chem. Soc., **98** (1976) 2985-2990.

AUTHOR INDEX

Reference numbers are **boldface**, chapter number first.
Page numbers for each reference follow in parenthesis.

SUBJECT INDEX